Copernicus Books

Sparking Curiosity and Explaining the World

Drawing inspiration from their Renaissance namesake, Copernicus books revolve around scientific curiosity and discovery. Authored by experts from around the world, our books strive to break down barriers and make scientific knowledge more accessible to the public, tackling modern concepts and technologies in a nontechnical and engaging way. Copernicus books are always written with the lay reader in mind, offering introductory forays into different fields to show how the world of science is transforming our daily lives. From astronomy to medicine, business to biology, you will find herein an enriching collection of literature that answers your questions and inspires you to ask even more.

Michael Yampolsky

The End(s) of Knowledge

 Springer

Michael Yampolsky
Department of Mathematics
University of Toronto
Toronto, ON, Canada

ISSN 2731-8982   ISSN 2731-8990   (electronic)
Copernicus Books
ISBN 978-3-032-01422-1   ISBN 978-3-032-01423-8   (eBook)
https://doi.org/10.1007/978-3-032-01423-8

Preface

<blockquote>
To know that we know what we know, and to know that we do not know what we do not know, that is true knowledge

Nicolaus Copernicus
</blockquote>

A few years ago I came to a weird realization about my research. To put things in context, I am a professional mathematician, and most of the time I work out puzzles of "pure math". When I tell this to strangers, the second most common response (the first being awkward silence) is "can you explain what you do in simple terms"? Well, the funny thing is that doing just that I have noticed that my research interests coalesce around a single unifying theme. That, in itself, is not necessarily surprising. A researcher's toolbox is limited, "if all you have is a hammer, everything looks like a nail", yadda, yadda. The surprising part was the theme itself. Without realizing this, I have been working on understanding the *limits of understanding*. Not some challenges or obstacles that can be overcome. No. The hard mathematical boundaries, which preclude any form of knowing, understanding, learning, or whatever we call making sense of the world beyond them.

This sort of thinking is foreign to the can-do spirit of modern science, which seems to be that with enough effort and grant funding we can figure anything out. Yes, there are some limitations to human abilities – we cannot fly, for example, but we have found scientific ways around that.

A similar spirit exists in mathematics. An easy to explain example is finding roots of (polynomial) equations. It is a ridiculously important skill, one, without which most of modern science would not be possible. School children around the world are made to memorize the "quadratic formula" for solving a quadratic equation. There are similar, although much more cumbersome to apply, formulas for solving equations of degree three and four. Published in a revolutionary treatise on algebra, the *Ars Magna*, in 1545 by Gerolamo Cardano, they are some of the greatest scientific achievements of the Renaissance. The history of their discovery is full of drama, replete with secrecy, deceit, and priority disputes. Yet, they represent a dead end of human thought. In 1824, Niels Henrik Abel proved the eponymous Abel's Impossibility Theorem: there are no algebraic formulas for solving equations of degree five or higher. In fact, it is possible to present explicit examples of equations whose roots cannot be written as algebraic expressions:

$$x^5 - x - 1 = 0$$

is one of them. If there were a duke or a prince in the 1600s who would endow a mathematical institute for solving polynomial equations, and would bestow salaries and grants on mathematicians toiling on this problem, it would not change the fact that its solution simply does not exist – it is a hard boundary. Throwing bodies of researchers on it would not help.

But, in the familiar can-do spirit, a mathematical workaround exists in the realm of analysis rather than algebra. Even though there are no formulas, there are numerous methods for *estimating* the values of the roots, with any desired precision. None is more famous than Newton's method, which originated in the works of Isaac Newton himself in the 1660s, and has taken the now standard form, found in Calculus textbooks, in the work of one Thomas Simpson in 1740.

There is something different about the boundaries I am going to describe, as they stand in the way of scientific research itself. Time and again we see scientific debates turn into political fights, decided in the court of public opinion more as matters of faith than science. This is nothing new, of course, as Galileo or Copernicus could have attested. Yet it is seemingly becoming more common even on trifling matters.

For instance, researchers have long insisted that moderate consumption of red wine brings health benefits. Many studies seem to confirm it conclusively. Take, for instance, the interview with a researcher from a 2015 *Time* magazine article, who gushed: "really solid gold-standard kind of research confirms our intuitions about the beneficial effects of moderate alcohol intake." Yet, a few months ago, the Canadian government radically changed its guidance, stating, in a firmly puritanical tone, that "no amount of alcohol is safe to consume". Based on the "science", no less. As a wine aficionado, I know what science I would choose to believe. But, all jokes aside, should it not be possible to settle this little matter conclusively? Once and for all?

Not likely. Statistical studies come with a sea of confounding factors, such as health, age, exercise habits, and preference for wines from Bordeaux versus from Bourgogne, to randomly name a few. Any possible statistical signal from wine drinking is drowned out by this noise. Studies with small samples suffer from all sorts of biases and could show drastically different effects for different subgroups of the population. All of this is generously sprinkled with researchers' own biases.

As I will hopefully explain to you, in an extreme scenario there may be precise mathematical reasons which make scientific understanding literally impossible. There is no going around them – they are the end of all knowledge. Yet, without realizing it, researchers may keep looking for understanding where none exists, finding themselves in ever-increasing confusion. In the words of famous scholar of human (and scientific) behavior and Nobel laureate Daniel Kahneman, "We're blind to our blindness."

This may translate into public confusion and loud political debates – mostly harmless in the case of red wine, but quite serious in other scientific matters. Once the boundaries of knowledge are reached, more work and more scientific minds thrown into the breach would not help, just as they would not help with searching for root-finding formulas which do not exist.

The idea that there may be a boundary beyond which knowledge ends is not new, of course, and has found its way into popular culture. Recently, reading the sci-fi novel "The Three-Body Problem" by Chinese writer Liu Cixin (which inspired a *Netflix* series of the same name) I was amused to see a description of a fictional group of scientists working "to use the methods of science to discover the limits of science, to try to find out if there is a limit of how deeply and precisely science can know nature – a boundary beyond which science cannot go."[1] Well, count me in as a member.

In trying to summarize where knowledge ends, one is inevitably led to think how lucky humans have been *so far* at understanding the world, at both a conscious and an unconscious level. The seemingly meaningless confusion of the universe can drive a scientist mad. Apocryphally, the great Aristotle so despaired to resolve the mystery of the ebb and flow of the tides, that he drowned himself in a tidal wave. Have a look at a four-week tide chart[2] for Chalcis, the place of his death, to appreciate the difficulty of the riddle he succumbed to:

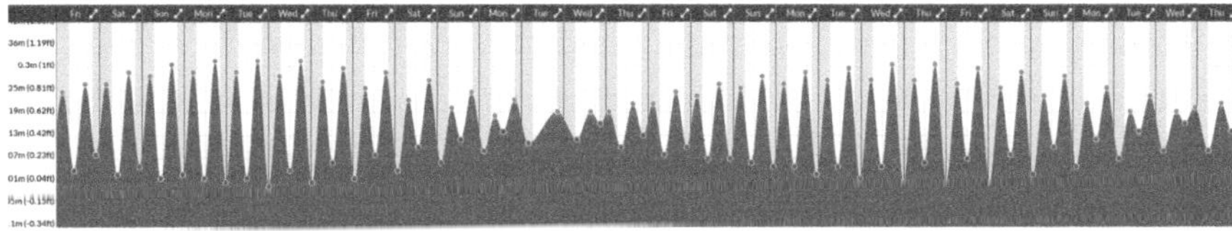

This story would have been an apt metaphor for science drowning in confusing data, yet a concise explanation of the periodic change of the sea level – the gravitational pull of the Moon – was there to be found by Isaac Newton. The overwhelming complexity of Nature yielded a tidy mathematical law of the tides.

As we will see repeatedly in this book, such triumphs of science are not guaranteed. We may be – and very likely already are – facing physical

---

[1] The exact source of scientific confusion featured in the novel – chaotic dynamics of which the three-body problem is an example – is simultaneously more and less understandable than the author believed. I will talk about it at length further on.

[2] Generated by *www.tide-forecast.com*.

phenomena which can be neither explained nor understood. That said, the boundaries of knowledge must have been very generously drawn, to afford our scientific understanding of Nature ample room to develop.

In a less generous universe, researchers would still be staring at the tides with no hope of ever discovering the theory behind them.

Many a great scientific mind of the past has felt that we have been unreasonably and undeservedly fortunate in this. Isaac Newton saw the answer in religion: "Don't doubt the Creator because it is inconceivable that accidents alone could be the controller of this universe". Albert Einstein was of a similar belief: "Every one who is seriously involved in the pursuit of science becomes convinced that a spirit is manifest in the laws of the Universe - a spirit vastly superior to that of man..."

I will not dwell upon such contentious matters. My goal is simply to convince you that there is a wall beyond which we cannot see and that our minds bump into it again and again. The mathematical principles on which this wall is built are not going to be overcome by brute scientific force, grant money, or breakthroughs in artificial intelligence. Invisible to us, in some cases, this wall may have already been reached.

I did not mean this book to be a dire warning about the fate of science, or even to be serious writing. I had fun summarizing my thoughts and trying to flesh out some common topics that I now realize motivate my own research and that of many of my colleagues. I allowed myself to play with other people's words in the form of witty epigraphs, apologies if you find it annoying. And I am fascinated with the ability of AI chatbots to draw, so I have spent perhaps too much time getting them to draw whimsical illustrations. Ditto. There is not too much math in the book, and one does not need to be mathematically gifted to read it – but it is a book *about* math, so formulas will pop up every now and then. Having said all this, I hope that you will find some intellectual amusement in this book – I did.

Acknowledgements and apologies

I am grateful to the friends and colleagues who have read preliminary drafts of this book and offered their advice and encouragement. Though I can be stubborn and have resisted some of the suggestions, their input has shaped my thinking and sharpened my writing.

I am especially thankful to my friend and academic collaborator, Cristobal Rojas, for his careful reading and critique of an early version of this book, as well as our many insightful discussions on the nature of knowledge.

I owe a great debt of gratitude to my dear friend Leslie Boctor, who generously donated her time and editorial expertise to help beat my text into shape.

Now, to the apologies. I am familiar with the exasperation of seeing loose interpretations of my own work, so I wish to preemptively apologize to my colleagues whose work I have quoted. I aimed for clarity without compromising too much precision – and I am sorry if I occasionally fell short. I include many historical anecdotes, but I must acknowledge that I am not a historian of science. To those historians I may have inadvertently annoyed, I offer my apologies.

I hope any errors you encounter will not detract from my main argument. And, in a general appeal to your generosity, I can only echo the immortal plea: "Please don't shoot at the pianist; he is doing his best."

Contents

Chapter 1.   Lost in the higher dimensional woods   1

One-dimensional thinkers   2

Maps and string   4

Of donuts and coffee mugs   7

A knot or not?   13

Whaddaya mean, doesn't exist?!   15

Chapter 2.   The most incomprehensible thing   19

Celestial clockwork   20

Separating the wheat from the chaff   30

Much shouting   33

Chapter 3.   The noise and the signal   37

The science of fitness   39

Fit or not?   41

Kolmogorov's razor   44

Chapter 4.   Live and learn   51

Fractals   54

A teachable moment about learning   56

Presenting the evidence   58

Playing chess against an alarm clock   60

Deep learning   62

Chapter 5.   Out of order, chaos   67
   A butterfly flaps its wings   68
   The King of Sweden's prize   72
   Of lynx and hares   78

Chapter 6.   Playing dice with the Universe   83
   Rolling the dice   85
   Losing at Monte Carlo   90

TL;DR   99

Bibliography   103

Chapter 1

Lost in the higher dimensional woods

> Without geography you're nowhere
>
> ———
>
> Jimmy Buffett

You may be thinking that the "boundaries of science" whose existence I grandly proclaimed in the introduction refer to some exotic, obscure-sounding scientfic or mathematical problems. Something that you would never notice or care about in your daily life, some technobabble. Quite the contrary. Take the limit of knowledge which I am about to describe in this chapter. It is so well familiar to all of us that we have largely stopped noticing it and no longer wonder *why it is there*. Just what it is, I will tell you in the next paragraph. The rest of the chapter is there to explain why it is an impenetrable barrier.

M. Yampolsky, *The End(s) of Knowledge*, Copernicus Books,
https://doi.org/10.1007/978-3-032-01423-8_1

One-dimensional thinkers

> You are a victim of your own neural architecture which doesn't permit you to imagine anything outside of three dimensions. Even two dimensions. People know they can't visualise four or five dimensions, but they think they can close their eyes and see two dimensions. But they can't.
>
> Leonard Susskind, a leading string theorist

> For string theory to make sense, the universe should have nine spacial dimensions and one time dimension, for a total of ten dimensions.
>
> Brian Greene, a leading string theorist

I am unable to comprehend four-dimensional objects. I could not form a convincing picture of one in my mind. I am not alone in this. In fact, I have never met anyone who could, even though a lot of my acquaintances are professional mathematicians and physicists who study multi-dimensional spaces for a living. You could say that there is nothing particularly surprising here – after all, we live in a three dimensional universe, so there is no real *need* for me to visualize anything in more than three dimensions. Maybe. However, it is just as possible that our belief in the three-dimensionality of the universe is shaped by the limitations of our own perception. Desert ants, as it turns out [5], believe that the world is two-dimensional – how silly of the desert ants. Physicists routinely propose multi-dimensional models for the Universe, describing the additional dimensions as "hidden" – that is, hidden from human comprehension. May the reason why visualizing four-dimensional objects is hard have at least something to do with the limits of understanding? Could it be something not even possible? Not just for a human brain, but not possible in principle?

Ok, no doubt you think that this is too far-fetched, so let me take a small step back, and try to convince you that thinking in higher dimensions is *hard*.

Here is a provocative question for you to consider: why are medical doctors one-dimensional thinkers? Let me explain myself, before my family doctor bans me from her office. A medical diagnosis can use a number of measurements, starting with the most basic ones, such as body temperature, or blood pressure. Body temperature can be, for instance, elevated or not, blood pressure can be in the normal range, or too high, or too low, and so on. When considered separately, these are single-dimensional measurements. A true two-dimensional diagnostics would use two measurements *simultaneously* and make conclusions based on both of their values at the same time. This may look like a "landscape", such as the one pictured in the following figure with peaks corresponding, for instance, to worse health outcomes, and valleys to better ones. In fact, the landscape could look more elaborate, with some parts of it rolling over the others, creating a more complex two-dimensional surface. Of course, such a diagnostic technique would be much harder to learn. But would it not be worth it, as it would be more accurate? Medicine is a highly competitive field, so why have doctors not evolved the ability of to do this over the course of the centuries?

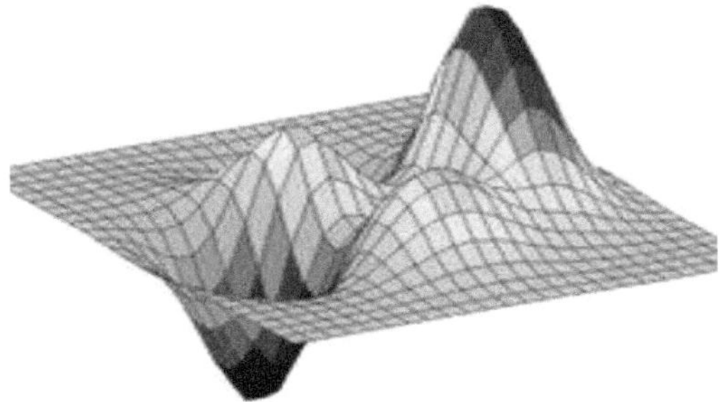

A simple two-dimensional landscape.

As we will see, understanding higher-dimensional landscapes may not only be too hard and confusing to be worth the effort, but, indeed, unattainable if their dimensionality is high enough.

Maps and string

> Carry with you maps and string, flashlights, friends who make you sing, and stars to help you find your place...
>
> — Mary Chapin Carpenter

> Maps are a way of organizing wonder
>
> — Peter Steinhart, a writer and naturalist

> I am told there are people who do not care for maps, and I find it hard to believe
>
> — Robert Louis Stevenson

To wander around a strange landscape without getting lost, we need a method of pinpointing our position. The ancient Greek hero Theseus is credited with slaying a monstrous half-man, half-bull, Minotaur, who was living at the center of the Labyrinth – an elaborate maze. To be able to find his way out of the Labyrinth, Theseus followed the string given to him by princess Ariadne, which he tied to the door post as he entered the maze, and kept unrolling as he walked.

For all the complexity of the maze, a location inside of it could be identified by a single number: the length of the unrolled Ariadne's string to reach it. Numbers used to identify a position are called *coordinates*. As it took a single coordinate for Theseus to find his bearings, the Labyrinth was *one-dimensional*. And it is not that hard to not get lost in a one-dimensional world. This is the mathematical moral of Theseus' and Ariadne's story.

A description of the world using coordinates is what we commonly call a *map*. Ariadne's thread is a map of the one-dimensional Labyrinth. In mathematical terms, the key to comprehending the configuration of a

Theseus, on his way to exposing the hidden mathematical simplicity of the Labyrinth.

landscape lies in *mapping* it. Without a map, we would be lost – like those unfortunate brave souls who blindly wandered into the Labyrinth before Theseus. Even more importantly, a map allows us to recognize an object for what it is, uncovering the inner simplicity in a maze. Mapping is, in short, a mathematical way of understanding the world. So – to appreciate the difficulty of multi-dimensional thinking – let us talk about the business of making maps.

To ground ourselves in the craft of mapmaking let us start with a familiar example: the GPS sensor in my phone uses two angles to anchor its position on the surface of the Earth. As most of us would at least vaguely recall, a circle drawn on the Earth's globe through the North and South poles is called a meridian. The prime meridian (or the Prime Meridian, as it is often ostentatiously capitalized) passes through the Royal Greenwich observatory (Observatory) in London. Any other meridian is obtained by rotating the Prime one by some angle which is known as the longitude. It is measured in degrees, from $-180°$ to $+180°$, increasing in the direction of the rotation of the Earth, from West to East. Since it takes 24 hours for our planet to make one full spin around its axis, and sweep $360°$ degrees of longitude, in one hour each meridian rotates 15 degrees. So with each 15 degrees increment of longitude, the sun rises over the horizon one hour earlier, which nicely ties this geographical measurement with the concept of local time. Watches have initially evolved as navigation instruments, to measure the change in the local time over the course of a sea voyage.

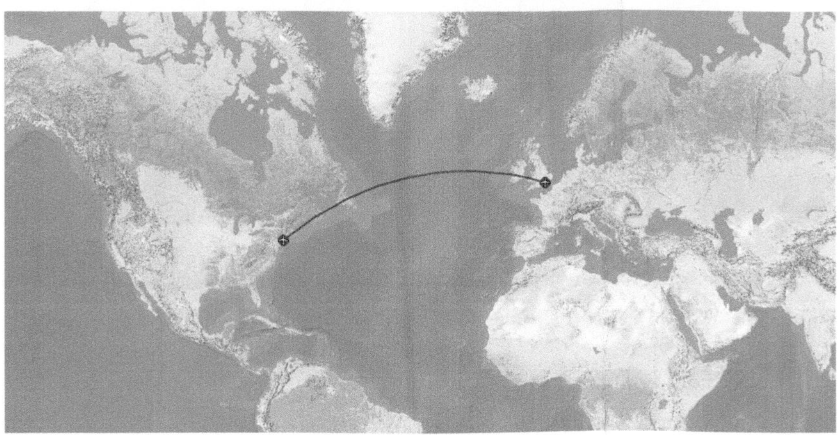

Shortest flight path from New York City to London (the arc of a great circle) in Mercator's projection, generated by GPS Visualizer, *www.gpsvisualizer.com*. It does **not** follow the rhumb line, which would look straight in Mercator's coordinates.

The other angle the GPS uses, the lattitude, measures the inclination over the plane through the center of the Earth which is perpendicular to its north-south axis of rotation. Since the North Star has been conveniently placed over the North Pole on the axis of the rotation of our planet, the seafarers of old in the Northern hemisphere could figure out the lattitude by simply measuring the angle at which it was seen at night. The ancient mariners in the Southern hemisphere were not as lucky, but made do with other, less conveniently located, stars.

The lattitude/longitude coordinate system on the Earth's surface is known as spherical coordinates. It is far from the only way of putting a preferred set of coordinates on the globe, and cartographers have made many other proposals over the course of history. Each of them offers some set of advantages. Most modern maps use some form of conformal coordinates – that is, coordinates in which traveling on the same course (the rhumb line in marine lingo) is represented by a straight line. The best known and most commonly used of these was invented by Gerardus Mercator in 1569, and bears his name. Meridians become vertical lines on a Mercator map, which creates a pretty obscene distortion of the map near the poles, but since most of us do not live near the poles, we rarely care.

Of donuts and coffee mugs

> The ancients had little doubt about the true shape of the earth: "It's [the world's] shape has the rounded appearance of a perfect sphere. This is shown first of all by the name of 'orb' which is bestowed upon it by the general consent of mankind. ...Our eyesight also confirms this belief, because the firmament presents the aspect of a concave hemisphere equidistant in every direction, which would be impossible in the case of any other figure."
>
> Pliny the Elder

No matter what cartographic system is used to map the Earth, one unfortunate obstacle it has to face is that the surface of the planet is not, actually, a sphere. Sometimes, the term *geoid* is used to describe a not-quite-spherical shape with indentations at the poles and a bulge along the equator created by the centrifugal force of the Earth's rotation, with a bit of gravitational pull thrown in. The mountains, valleys, hills, and skyscrapers are icing on the cake. In superimposing a coordinate system, cartographers have to smooth and round the planet's shape, as if it were made from play-doh. Only then do they get a neatly spherical globe on which meridians and parallels fit. This excercise is a feature, not a bug, of mapping.

The crumpled sheet of paper may look complicated but deep down it is still only two dimensional, and can be uncrumpled to a flat surface and mapped using a coordinate grid. A small bug would not even know that the sheet is crumpled and could follow the gridlines as if they were straight. Just as we follow parallels and meridians and not notice the mountains and canyons along the way.

A mapping technique only makes practical sense if it is the same for similar objects – that is, for the ones that can be reshaped into one another. The Moon is not shaped exactly like the Earth, but its shape can also be massaged into a sphere, so the same mapping techniques that work on the Earth are used to produce lunar atlases. To get a better feel for this, imagine yourself as a small insect (sorry) crawling on a sheet of graph paper, ruled with a fine grid of horizontal and vertical lines. Using these lines as a coordinate grid, you can answer practical questions, such as in what direction you would need to crawl to reach the nearest edge of the sheet, or how long it would take you to reach this edge. If the sheet is crumpled, the same coordinate grid would give you the same answers.

Mathematicians like to say that one object can be "deformed" into another to avoid using the terms like "massaging" or "uncrumpling". A donut can be deformed into a coffee mug. In technical speak, a donut and a mug are *homeomorphic*, from Greek *homoios* "like" and *morphē* "form". As we are about to see, the surface of a donut is pretty easy to map, which can be used to produce a convenient map of a coffee mug... well, convenient if you are a bug crawling over a coffee mug. The surface of the Moon, or of Mars, or of a solid brick, is homeomorphic to a two-dimensional sphere. An extreme example is the surface of the human brain. Its many folds and creases vary from one person to the next, making it difficult for neuroscientists (and neurosurgeons) to recognize the specific regions which correspond to, for instance, specific motor functions. A branch of modern applied mathematics deals with brain mapping – that is, deforming a brain surface into a sphere to produce a readable functional map of the brain.

Deforming a coffee mug into a donut back to a coffee mug, by Keenan Crane and Henry Segerman, reproduced with permission.

To know a multi-dimensional object is to map it. Paradoxically, in the early 2000s, two mathematicians, Alex Nabutovsky from the University of Toronto and Shmuel Weinberger from the University of Chicago, demonstrated that cartography is not generally possible for shapes of dimension five and higher (Weinberger has written a specialist book on this [21]). Let me repeat this, for greater emphasis: no general mapping techniques are

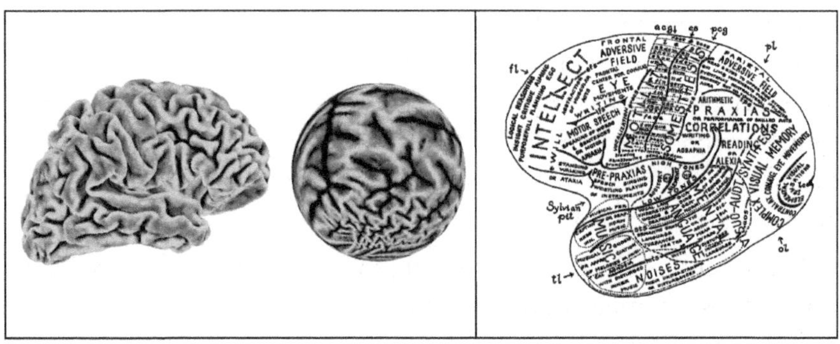

Left: mapping of a brain onto a spherical surface. Image by Gary Pui-Tung Choi, reproduced with permission.
Right: a somewhat cartoonish schematic functional map of the brain. From: Fig. 275, p. 456 of "The Vertebrate Visual System" by Stephen Polyak.

available for five-dimensional shapes. At all. And if this sounds weird, wait until I tell you why that is the case.

It turns out that the problem goes somewhat deeper than mapping: it lies in *recognizing* five-dimensional objects. I realize that this sounds vague, so let me go back to two-dimensional surfaces to explain what I mean.

Recognizing that two shapes are homeomorphic is key to mapping. The surface of a human brain can be mapped onto a sphere, the same way as the surface of the Earth or of the Moon. The same coordinate system can be put on the surface of a donut as on that of a coffee mug – flip the page and I will show you how. Once you have seen that a donut can be deformed into a coffee mug, and vice versa, you cannot unsee it. But let's be honest with ourselves, it would not have been the most obvious thing to notice. Geography relies on being able to tell when two surfaces are homeomorphic to each other. Is there a sure-fire way of doing that? It turns out that there is, and it is a thing of mathematical beauty. Let me tell you about it.

We think of a surface as having two sides – an "outside" and an "inside". A good visual illustration is the skin of an orange – bright, shiny orange on the outside, white and fibery on the inside. We cannot get from

A subterranian world.

the outside to the inside without breaking the skin. Let me call such surfaces *two-sided*, although this is not the terminology commonly used in mathematical literature. The Earth's surface is two-sided, so is the surface of a Helium-filled balloon, or the surface of a donut. A branch of "subterranian" science fiction imagines life on the inside surface of the hollow Earth. Jules Verne's "Journey to the Center of the Earth" is probably the best known example, its characters find subterranian life on the other side of the surface, only 87 miles deep. A Russian novel "Plutonia" by V. Obruchev imagines the inner surface lit by an inner Sun, located at the center of the Earth. Among many other oeuvres of the genre is the "Icosaméron" written in 1788 by Giacomo Casanova (yes, that Casanova).

If the Earth were flat, then one could conceivably travel from one side to the other over its edge, so it would not be truly two-sided. Neither would be the skin of an orange if we make a hole in it, as it is again possible to connect the inside with the outside over the edge of the hole. A Möbius band is a famous example of a one-sided surface on which you do not need to approach the edge to travel from "outside" to "inside".

A ride on a Möbius rollercoaster.

A great disovery of early 20s century Mathematics is that every two-sided surface is homeomorhic to a donut with g donutholes. The number g (for *genus*) can be zero, a donut with zero holes is a spherical cake, whose surface is homeomorphic to a sphere. If $g = 1$, then we get a familiar donut shape; $g = 2$ is a donut with an extra hole; $g = 3$ is the surface of a pretzel, and so on. Such a powerful classification makes it easy for a human brain, or for an artificial one, to *recognize* two-dimensional shapes – you only need to count the donutholes. That is how I knew that a coffee mug is homeomorphic to a single-holed donut!

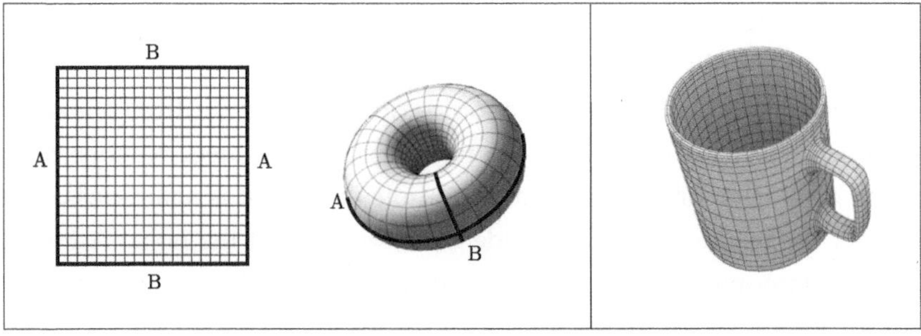

Left: to make a donut out of a sheet of graph paper, glue the opposite sides. Right: a coordinate grid on a coffee mug.

Now what about mapping two-dimensional shapes? We have already talked about mapping a sphere. Cartography on a donut with $g \geq 1$ holes is made easy by the fact that it can be obtained from a regular $4g$-sided polygon by gluing together pairs of sides. If you glue together the opposite sides of a square, the resulting shape will be a traditional single-holed donut. Putting a coordinate grid on the square, we get a coordinate system on the surface of the donut – and the same exercise can be done for a donut of any genus. Once we have recognized the genus of a two-dimensional shape, we can transfer the coordinates from the polygonal $4g$-sided map to it. A computer can be taught to do this, just as it can be taught to map the surface of a brain to a sphere.

A knot or not?

> Gordian knot, a knot impossible to unravel, and Alexander the Great solved the problem by cutting through it with his sword, in other words by cheating.
>
> Patricia Cornwell

> I am a Gordian knot. Don't unravel, just slice.
>
> Orson Scott Card

Okay, so to map a two-sided shape we need to identify it as a donut with g donutholes – once that is done, there is a general method of mapping it. However, recognizing a shape is not always a straightforward exercise even if you have some prior idea of what it may be. An instructive example is given by a branch of Mathematics known as Knot Theory which struggles with a seemingly simpler problem of identifying shapes. For a mathematician, a *knot* is a continuous loop of string, without loose ends. An "unknot" is a simple loop, looking like a circle made of string. One cannot untie a non-trivial knot and turn it into an unknot without cutting the string. But twisting and turning the string can make the knot look far more complicated than it is. If an unknot is tangled badly enough, then it

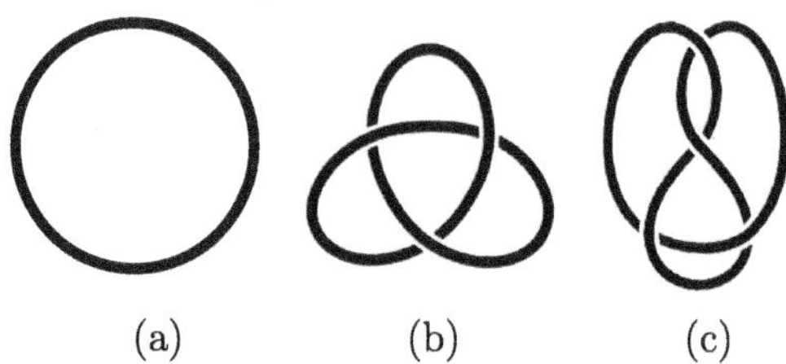

Some knots: (a) an unknot, (b) a trefoil knot, (c) a figure-eight knot.

will be hard to recognize that it is, in fact, an unknot, and to untangle it –
as anyone who has ever tried to straighten a garden hose can readily attest.

In figure (a) below you can see a prettily tangled unknot designed by
the German mathematician Lebrecht Goritz in the early 20s century. Do
you think you could untangle it? If that was not enough of a challenge
for you, have a look at the monstrosity in figure (b). It was designed by
Wolfgang Haken, and is known as a Gordian unknot.

In 333BC Alexander the Great was challenged to untie an intricate knot
in the city of Gordium. The local legend said that whoever was able to
undo it would rule all of Asia. Alexander then cut the knot with his sword
(which seems consistent with Alexander's general approach to challenges).
Haken's Gordian unknot is not actually a knot, but it is tangled very badly.
Is there a systematic way to untangle a knot, to see if it is an unknot or not
(or knot)?

In 1954, the great Alan Turing (whose tragic life story was recently popularized in the movie "The Imitation Game") complained in his final published paper [18] "No systematic method is yet known by which one can
tell whether two knots are the same."

The first algorithm to determine whether a tangled loop of string is an
unknot was designed by the same Wolfgang Haken in 1961. It can be implemented on a computer, but is very slow for very tangled knots. The
Universe would probably end before Haken's algorithm would verify that
his Gordian unknot is indeed an unknot. As recently as 2021, the Oxford

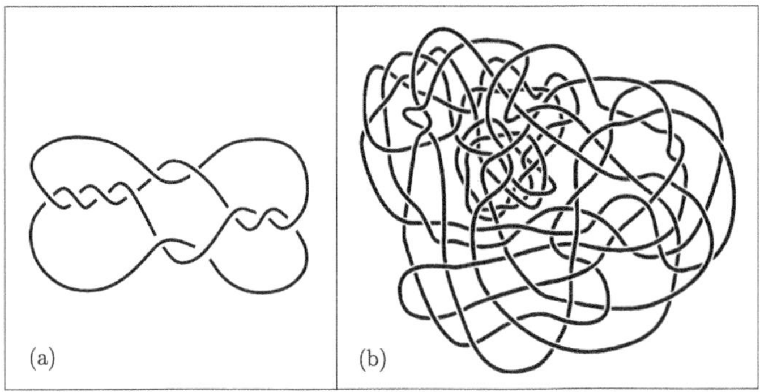

(a) a Goritz unknot. (b) Haken's Gordian unknot.

mathematician Marc Lackenby designed the currently optimal known algorithm, whose running time grows slightly faster than a power of n, where n is the number of times the loop in the picture crosses over itself.[1]

Whaddaya mean, doesn't exist?!

> ...in mathematics you don't understand things. You just get used to them.
>
> ———
>
> John von Neumann, a Hungarian-American polymath, a founder, among many other things of: Game Theory, Computer Science, the mathematical theory of Quantum Physics, and Operator Theory.

Now that we have seen how recognition of shapes is related to mapping, and how it can be an interesting and challenging problem even for basic shapes, let us go back to the result of Nabutovsky and Weinberger for five-dimensional shapes. It is based on the startling fact that there does not exist an algorithm for recognizing five-dimensional shapes.

---

[1] To be precise, the running time is bounded by $Bn^{C \log n}$, where B and C are some constants. This could still take a while for something like Haken's Gordian unknot, which crosses over itself a bunch of times before closing the loop, particularly if C is large. But computer scientists typically think of computational problems whose running time is bounded by a constant times a power of the length of their description as being computationally "easy". From this viewpoint, Lackenby's result means that the problem of deciding whether a knot is an unknot is "almost easy".

And by "does not exist" I do not mean "currently" – I mean that no such algorithm can ever be designed. It is not possible. This is a mathematical fact – a theorem.

The first ever example of a theorem of this sort was given by none else but Alan Turing in the year 1936 in a now famous paper [**19**]. It concerned a problem well-known to coding practitioners. There is that happy moment when your brand-new piece of code is finally debugged, and runs problem-free, without crashing or giving you errors. You look on happily as it runs, and runs, and runs, and runs, and runs... and as time goes on and it keeps on running, you begin to wonder whether it is still doing something useful or if it has gone into an infinite loop and will never halt. Turing called this dilemma the Halting Problem and gave a formal proof that no algorithm to solve The Halting Problem exists. The proof is taught in Theoretical Computer Science 101, but requires some specialized language – Turing had to carefully define what an algorithm or a computer program is, before ruling out its existence. But the intuitive idea is easy to explain. Suppose, by way of contradiction, that such an algorithm can be found. Let us implement it in a computer that we will call the Halting Checker. The Halting Checker can be shown the text of a program and will correctly answer "yes it halts" or "no it does not halt". Consider the following program:

```
If the Halting Checker says that this program halts,
    then run forever (go into an infinite loop).
If the Halting Checker says that this program does not halt,
    then halt.
```

This program by design will do the opposite of what the Halting Checker thinks that it does. And yet, the Halting Checker was supposed to only give correct answers. The only way out of this logical conundrum is to accept that the Halting Checker cannot exist – that is, there is no algorithm to solve the Halting Problem.

It is, of course, much harder to prove that there is no algorithm to recognize five-dimensional shapes (although, amusingly, it relies on a version of Turing's proof at its core). One can take a few pieces of five-dimensinal lego, put them together – and neither a computer nor a human brain would be able to tell if the resulting shape is homeomorphic to, for instance, the five-dimensional sphere.[2]

And if you cannot recognize it, then you would not be able to map it. Cartography is dead on arrival in five-dimensional spaces. If we need to find our way in a pile of data which requires at least five parameters to properly describe, then we are well and truly screwed. We are going to get lost.

What about lower dimensions? We are in good shape in three dimensional spaces: they can be mapped. A mathematical proof of this was given in a rather dramatic fashion by an eccentric mathematical genius from Russia by the name of Grigori Perelman in 2003. The book "Poincaré's

---

[2]The five-dimensional sphere consists of all points in the six-dimensional space whose distance to the origin is the same (for instance, equal to one).

Prize: The Hundred-Year Quest to Solve One of Math's Greatest Puzzles" by G. Szpiro describes some of the drama. An exhaustive list of possible geographies of three-dimensional objects was suggested by a giant of American Mathematics of the 20th century, William Thurston, and was known as Thurston's Geometrization Conjecture. Perelman's proof of the Conjecture followed an approach suggested by Richard Hamilton, a mathematician from Columbia University. Needless to say, the problem of mapping a three-dimensional space is not nearly as straightforward as its two-dimensional analogue, but at least it can be resolved.

When it comes to four-dimensional objects, the situation is presently less clear. Some of them cannot be mapped, for others, a hope of finding a cartographic description still remains.

In summary of this chapter, it would make a good sense for a doctor to employ two-dimensional diagnostic techniques. Mapping two-dimensional landscapes is not that hard. Three-dimensional diagnostics is generally too hard for a human being, but is possible, in principle – a clever AI bot could prove very useful here. Four-dimensional mapping *may* be impossible, and five-dimensional mapping is impossible for sure. Our cognition is constrained by these mathematical limitations, so it is, perhaps, not that surprising that we cannot think outside of a three-dimensional "box".

Chapter 2

The most incomprehensible thing

> The most incomprehensible
> thing about the universe is that
> it is comprehensible
>
> Albert Einstein

Seeing that you have kept on reading, I hope I have convinced you, or at least made a plausible case, that our comprehension is limited to low dimensions. Phenomena which require more than a few (very few indeed) parameters to describe them may forever remain beyond our understanding. Such is hard mathematical reality, which tells us that, in a way, understanding in many dimensions *does not exist.* We obviously comprehend a great deal about the world around us. This means that we are able to reduce much of its complexity to descriptions using only a few "coordinates". I wish I could tell you *why* that is possible. The *why* may indeed be the most incomprehensible thing about the universe. Instead, in this chapter I will focus on *how* such comprehension emerges. I will start with a very familiar example – one of the most important examples of scientific understanding in history. I will then try to formulate a general mathematical principle to see where the limit of understanding lies, and whether – a scary thought – in some instances, science may have already reached it.

© The Author(s), under exclusive license to Springer Nature Switzerland AG 2025
M. Yampolsky, *The End(s) of Knowledge*, Copernicus Books,
https://doi.org/10.1007/978-3-032-01423-8_2

Celestial clockwork

> The world embarrasses me, and
> I cannot dream that this watch
> exists and has no watchmaker.
>
> Voltaire

> The only watchmaker is the
> blind forces of physics.
>
> Richard Dawkins

> If only I had known, I should
> have become a watchmaker.
>
> Albert Einstein

The world around us is infinitely complex – by rights, we should forever remain confused about its workings. Historical examples of such confusion abound. None is more instructive than the multi-millenial error-filled quest to describe the machinery of the Solar system. I very highly recommend the classical, aptly named book by Arthur Koestler "The Sleepwalkers" [8] on the subject, the sleepwalkers being the scientists trying to comprehend the rules of planetary motion. There obviously *is* something to comprehend – the motion of celestial bodies in the Solar system seems to follow cyclical patterns, repeating after regular intervals of time. In the modern scientific terminology, such cyclical motion is called *periodic*. Some examples of this periodicity, such as the changes of the phase of the Moon, or of the length of the day are apparent to all of us. Others, such as planetary conjunctions or eclipses, emerged readily from the earliest astronomical records. And yet, the search for an understanding of what we now think of as the "laws" of physics responsible was mired in confusion for millenia. And the strangest thing about this search is that the laws were actually there to be found.

Imagine that you are handed a complicated mechanical clockwork, showing a number of planets, the Sun, some other stars, and several prominent constellations, and are asked to figure out its design without opening it up and looking at the mechanism. If there were only a couple of celestial bodies involved, you would probably stand a chance. But this clockwork is so involved, with each part performing its own endlessly repeating cycle of movement, that, after a while, you mostly give up trying and start to formulate random guesses which might somewhat describe the reality, and are heavily influenced by how you would *want* things to be running, rather than how they actually are.

This is, basically, where things were for centuries. The celebrated moments in antiquity when someone would get something right (most notably, that the Earth revolves around the Sun and not the other way around) seem more like accidents than systematic progress of science. Indeed, they would almost immediately be forgotten in favor of theories asserting the opposite, or would provoke fierce theological disputes and be dismissed based on nothing – but then, they were themselves based on nothing, so would be easy to dismiss.

Until, that is, a successful, truly scientific model of the Solar system appeared, which stood the test of time. I am talking, of course, about the model designed by a genius of ancient thought, Claudius Ptolemy. Despite what you may have heard from your high school teacher, Ptolemy's model was scientific in the modern sense, that is, based on universal principles and agreeing with astronomical observations. Indeed, it agreed with them so well, that, after Ptolemy described it in his oeuvre *Almagest* circa 141AD, it remained relevant for nearly 1500 years. Its design principles were so sound that a version of it could have remained in use to this day. Rather than being a hindrance, it *inspired* the work of Copernicus. And yet, it (along with Copernicus' model) represented a dead end of science. But not for the reasons that would ever be discussed in an Astronomy class.

First, credit where it is due. It is easy to underestimate the difficulty of understanding the relative motion of the Earth, the Sun, and the Moon to the degree required to produce reliable Solar and Lunar calendars, for example. Throw in a few more planets, and the problem becomes truly staggering in its complexity. Ancient astronomers acted as historians, rather than scientists – recording previous cosmic events with the hope of guessing their repeating patterns. How easy was that? Search up the list of Lunar or Solar eclipses visible from your area/city/town (at the time of writing, *timeanddate.com* does an excellent job of it) for a decade or two and see if you could discern a pattern. Any pattern. Oh, and for an ancient astronomer, the task would itself be compromised by the lack of a good time-keeping mechanism (that is, an accurate calendar).

Against this background, Ptolemy's model was a breath of fresh mathematical air. It gave a satisfactory description to the mystery of the known universe, no more, no less. Moreover, the Ptolemy-designed clockwork came with a fine-tuning kit: it was easy to improve it based on the same mathematical principles as new data would come in – and improved it was. Before taking a dive into its workings, let me address the main criticism that lazy schoolbooks mount against Ptolemy – his model was Geo-, that is, Earth-centric, with our planet put in the center of the known Universe as an immovable object. However, in itself, this is hardly a worthy criticism. We live on the Earth after all, so any astronomical model should result in a description of movement *relative* to our position on this planet. And measuring things relative to our position in the Universe, by definition, makes our position immovable. Paradoxically, even starting with the Helio- or Sun-centric model, we will end up producing Geo-centric calendars of cosmic events, because that is what interests us in the end. So, from that point

of view, making everything revolve around the Earth from the outset may
not be a bad idea.

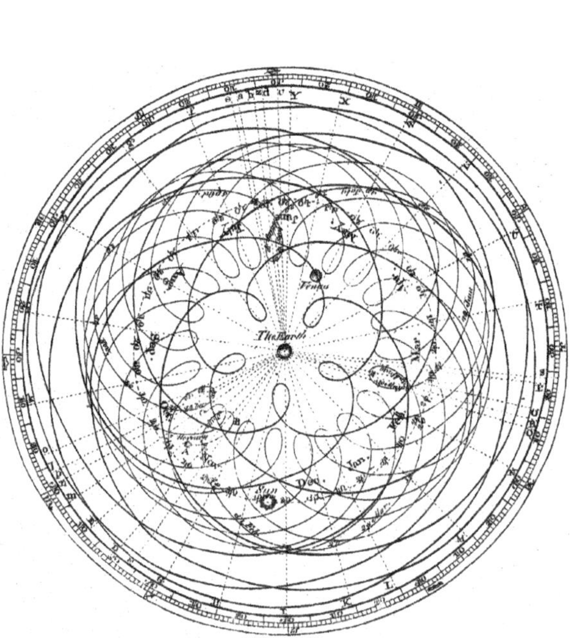

Representation of the apparent motion of the Sun, Mercury, and Venus
from the Earth. Taken from the "Astronomy" article in the first edition of
Encyclopædia Britannica (1771), drawn by James Ferguson (1710-1776),
based on similar diagrams by Giovanni Cassini (1625-1712). This geo-
centric diagram shows, from the location of the Earth, the Sun's apparent
annual orbit, the orbit of Mercury for 7 years, and the orbit of Venus for
8 years, after which Venus returns to almost the same apparent position
in relation to the Earth.

Ptolemy's work continued in the great tradition of ancient Greek mathematics and Babylonian, and then Greek, astronomy. In that tradition it
was assumed, in particular, that all of the complexity of cosmic patterns
could be reduced to one basic type of movement – uniformly in perfect circles[1]. This was almost a random guess, but powerful in its simplicity and,
as we will see, entirely mathematically adequate. It quite literally imagines
the Universe as a mechanical clockwork driven by an assembly of round

---

[1] In the modern telling, Aristotle usually gets the blame for this.

gears. At first go, the idea of uniform circular motion around the Earth is a terrible match to the mechanics of the Solar system. The observed motion of the planets is not circular, and the speed of their movement is not uniform. They speed up and slow down, and sometimes seemingly reverse course and go backwards (the phenomenon known as *being in retrograde* that astrologers tend to give particular significance to). And yet, things can be fixed by adding a uniform circular motion on top of another uniform circular motion.

Imagine a mechanical gearbox in which a large gear rotates uniformly around a fixed axis. Its teeth engage the teeth of a smaller gear, which both rotates uniformly around its center and moves around the gear that drives it. The combination of these two motions will produce something far more complex. If we mark a dot on the edge of the smaller gear, then its motion, once both gears have completed the turn, will produce a loop whose shape can be quite complicated and will very much depend on the ratio of the sizes of the two gears. Have a look at the next figure which pictures an aptly named *planetary gearbox* realizing this motion (for obvious technical reasons, it has three small gears rotating around the main central one, rather than just one.)

A planetary gearbox (source: Wikimedia commons).

And that's just with two gears – if we add three or more, then the loops could become more and more elaborate. The fantastic discovery of Ptolemy was that he could accurately reproduce the motion of all known planets except one by placing it on the edge of a smaller gear, driven by a larger gear, whose center would be nearly, but not exactly, the Earth. To complicate matters a bit more, the motion of the smaller gear around the larger one would be uniform with respect to some offset point – and not the center of the larger gear[2]. The two uniform circular motions would then combine to produce the observed non-uniform yet periodic orbit of the planet. And what about the single exception? That planet, Mercury, required an extra gear – three in total. Rather than serving as a disappointment, it was a mark of the success of Ptolemy's approach – want to match a more compicated orbit? Just add more gears.

Now, all of this may seem like a rather unexpected stroke of luck. But let me tell you that there is a whole branch of Mathematics known as Fourier analysis which studies the periodic motions obtained in this manner. Its fundamental fact is that **any periodic motion can be approximated closely enough as to be indistinguishable by selecting a finite number of appropriate gears** (look at the figure below for an illustration). Mathematically, such combinations are expressed by *Fourier sums*. The motion of a celestial body in Ptolemy's original model is not, strictly speaking, described by a Fourier sum, but by a slightly more complicated mathematical construct. Arrangements of gears which represented actual Fourier sums were introduced in his 1543 text *De revolutionibus orbium coelestium* by none other than Copernicus, whose model was, of course, Helio-centric. Unknown at the time, this fact made Ptolemy's approach a universal tool for describing the Solar system or, indeed, the observable Universe. By adding more and more gears and fine tuning their sizes, all experimental discrepancies could be accounted for. The workings of the Solar System could thus be recreated as an actual mechanical clockwork with round spinning gears.

---

[2]Ptolemy could have used an extra gear instead of introducing this offset point. He was conscious of this, and as a true watchmaker, made the choice in favor of having fewer gears.

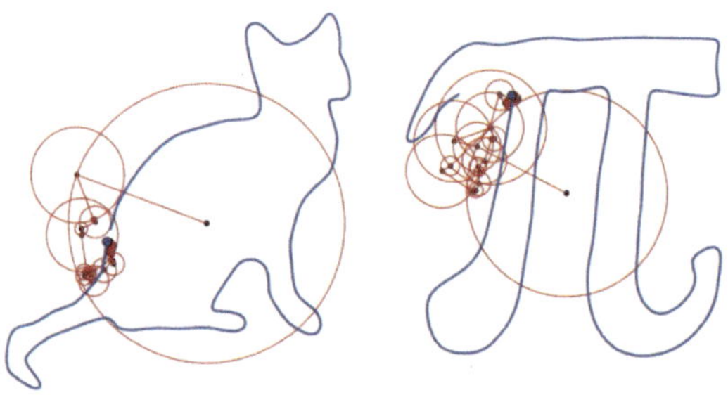

Tracing complicated loops with an arrangement of 63 gears. By Juan Carlos Ponce Campuzano from *www.dynamicmath.xyz*; reproduced with permission.

Why hasn't the story ended with Ptolemy? What ultimately drove Copernicus to question Ptolemy's model, and to come up with his own one, which was based on similar principles? What was the problem Copernicus attempted, and ultimately failed, to solve?

At its heart, this was not a theological dispute about the divine design of the Universe. Ptolemy himself was quite clear that his was only a mathematical model of the Solar system, and that, in particular, he could have used different arrangements of gears in its design. He did not claim that it was somehow sacred or irreplaceable, **only that it worked well with the astronomical data** available at that time. However, as experimental tools became better and measurements got more precise it needed adjustments. More gears! Copernicus attempted to resolve this increase in complexity. Indeed, making the Sun rather than the Earth the center of rotations allowed him to lower the number of gears required to explain the motion of each planet. Mathematically, it streamlined the model, expressing the motion properly as a Fourier sum (which Copernicus, of course, knew nothing about). Now further adjustments could be made in a straightforward way, by adding more terms to the sum, which would correspond to an extra layer of gears in the clockwork. And further measurements would bring further discrepancies which would need to be fixed this way – more terms, more gears.

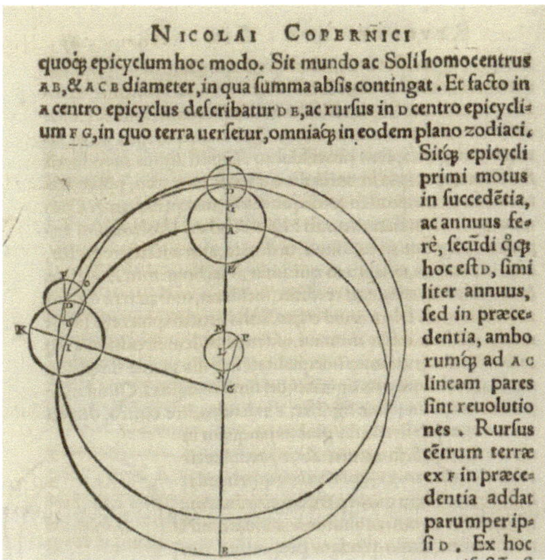

NICOLAI COPERNICI

quoᗅ epicyclum hoc modo. Sit mundo ac Soli homocentrus A B, & A C B diameter, in qua summa absis contingat. Et facto in A centro epicyclus describatur D B, ac rursus in D centro epicyclium F G, in quo terra uersetur, omniaᗅ in eodem plano zodiaci. Sitᗅ epicycli primi motus in succedētia, ac annuus ferè, secūdi ᗅᗅ hoc est D, similiter annuus, sed in præcedentia, amborumᗅ ad A C lineam pares sint reuolutiones. Rursus cētrum terræ ex F in præcedentia addat parumper ipsi D. Ex hoc

Copernicus' original model. Note, that all of the motions are perfectly circular.

Scientists instinctively recoil when new measurements require increasing the complexity of the model. They justifiably feel that as new variables (size and position of the new gears) need to be introduced, seemingly without end, the model's complexity will rival that of the Universe itself. We would trade the intractable periodic motion of the planets for the seemingly intractable workings of its mathematical representation. Just look at the left side of the next figure. It shows a Renaissance marvel: a clockwork built on Ptolemy's principles which accurately (within the measurement precision of its day) imitated celestial motions. Now, imagine a mechanical clockwork with many thousands of gears driving it that might have been needed today.

It is better, of course, to have such a descriptive model, than nothing at all. Predictions in either Ptolemy's model or Copernicus' one could be made on the basis of mathematical formulas. Astronomers were no longer being historians – they could foretell the future with accuracy. But had we gone down this road, there would be a limit on how much we could improve our understanding of the Solar system. Once our model had grown too complex for us to comprehend its own workings, we would

The motion of the Solar system as clockwork (known as an *orrery*). Left: a reconstruction of the first known orrery built in 1348 by Giovanni Dondi in Padova. This enormously complicated mechanical device attempts to implement the similarly enormously complicated Ptolemaic theories of planetary movement. Right: a 1766 Benjamin Martin orrery, used at Harvard. Source for both images: Wikimedia commons.

have reached the end of the road. There is a limit on how complex a mechanical clock we could actually build without losing track of the interaction of its moving parts. And when that limit would be reached, the design of the model would become an obstacle to further scientific progress as its complexity would overwhelm both scientific insight and computational resources. Science would hit the wall and fail. It would have come to an end.

And yet, a happy scientific ending arrived. In the early 1600s, Johannes Kepler proposed a much simpler conceptual model for the clockwork of the Solar system. It was designed on three principles, known now as Kepler's laws. The first two laws basically say that the clockwork of the Solar system is indeed arranged around the Sun, and the planets, including the Earth, revolve around it at a uniform rate[3]. Well, I am oversimplifying a bit. Kepler's laws say that the planets move in elliptical orbits, and that the arm of the clockwork from the Sun to the planet sweeps out equal areas over equal periods of time (going faster the closer the planet is to the Sun). But these ellipses are only very slightly non-round, so the planetary motion

---

[3]One exception is, of course, the Moon, whose orbit revolves around the Earth.

is very nearly uniformly circular. Kepler's third law is truly revolutionary (pardon the pun) in its mathematical precision and simplicity. Kepler himself was so impressed that he recorded the date of its discovery: 8 March 1618. This event could have been celebrated as the birthday of modern physics, and yet remains completely obscure: the only notable event on March 8th in all of the 1600s that Wikipedia currently knows about is the Roskilde peace treaty between Denmark and Sweden.

Without further digression, here is Kepler's third law: the *square* of the length of the period of revolution of a planet around the Sun – in other words, of that planet's Solar year – is proportional to the *cube* of the planet's distance to the Sun[4]. As an example, the distance from the Earth to the Sun is 149.6 million kilometers, for Venus it is 108.2 million kilometers. The ratio of the cubes of the distances is approximately 2.64. Kepler's law predicts that this is equal to the ratio of the squares of the Earth's and Venus' years, or, after taking square roots, that the period of rotation of the Earth around the Sun is about $\sqrt{2.64} \approx 1.62$ times longer than that of Venus. As we all know, the Earth's year has approximately 365.25 Earth days in it. The year on Venus has approximately 224.7 Earth days, which is, indeed about 1.62 times shorter.

Kepler's laws, in their remarkable simplicity, provide all the information you would need to design a clockwork which correctly describes the Solar system. To appreciate the contrast, look at the mechanical clockwork shown on the right side of the figure on page 28. It is a very simple device since, after Kepler, all that was needed were arms holding the planets rotating around the Sun with speeds agreeing with Kepler's third law (notably, this orrery does not reproduce the true elliptical shape of the orbits, to further emphasize simplicity at the expense of a slight inaccuracy). To me, the most intriguing part of this story is not the discovery itself – but that the simple mathematical rules were actually *there to be discovered* in the first place.

---

[4]Since the orbit is elliptical, the distance changes – but one can use, for instance, the maximal distance, or the minimal one, or the average one.

It is not difficult to imagine an alternative world in which the speed of the movement of a planet in the clockwork of the Solar system was significantly influenced by, say, the positions of all of the other planets, and by the positions of the constellations of the Zodiac seen from the Earth. Whether we used a Geo-centric model of Ptolemy or a Helio-centric model of Copernicus, such complexity would require us to use a clockwork with an ever increasing number of gears to describe its motion, without any prospect of a conceptual understanding. More precise measurements would only lead to a more complex description.

Oh, and of course, all of these things I mentioned do influence the speed and positions of the planets. As do solar flares, and the phases of the Moon. It is just that all of these parameters, as well as a myriad of other ones, do not make a substantial impact – the only variable that truly matters is the distance to the Sun. That is the type of miracle that Einstein was amazed by in the opening quote of this chapter.

Separating the wheat from the chaff

> Physics is becoming so unbelievably complex that it is taking longer and longer to train a physicist. It is taking so long, in fact, to train a physicist to the place where he understands the nature of physical problems that he is already too old to solve them.
>
> Eugene Paul Wigner, a Nobel Prize laureate in physics

In general, the difficulty in comprehending the universe is that there are just too many moving parts: from the particles forming all matter, to the leaves on the trees, to the stars and the galaxies. It is easy to imagine any attempt at understanding the world drowning in the sheer number of variables to take into account (quite poetically, physicists call these variables *degrees of freedom*). And yet, in a physical law of nature only a small number of degrees of freedom matters, the rest of them fades into irrelevance. So why do laws of physics *exist*? Why can simple mathematical formulas provide reliable predictions of seemingly infinitely complex natural phenomena?

By way of analogy, imagine a national electoral campaign. Each voter has a different set of concerns, interests, and worries – there is an almost infinite variety of opinions on all sort of things. However, typically, a rather small set of issues emerges to dominate the election. The voters polarize in their attitude towards these issues, the politicians do too – and so a few deciding election questions emerge from the seemingly endless list, to which the results of the election serve as the answers. This polarization on a limited set of issues is the defining feature of a functioning political system.

"...let's ignore the confusing details and focus on the main idea. We just need to find it in this pile of notes..."

Suppose, for example, that a campaign ends up being decided entirely on the question of gun control. It does not mean that each voter only cares about gun control. It does not even mean that most voters *mostly* care about gun control. But the one-dimensional scale of attitudes towards guns (more control versus less control) summarizes most of the polarization of the electorate. Any other political concern is largely irrelevant to the outcome: it is either very much tied to the attitude towards guns (and thus brings nothing new to the debate), or does not create a clear enough political signal to be worth pursuing.

Physical theories in which a few principal degrees of freedom emerge to dominate the narrative are known as *renormalizable*. Renormalization refers to a process of distilling and amplifying these degrees of freedom from the myriad other ones. It is a loaded term, and I apologize to the experts in the audience who may be unhappy with my oversimplified account. Let me attempt to explain renormalization, once again using the analogy with a national electoral campaign, in which crystallization of voters' intentions typically starts at a very local level, in a neighborhood, a town, an electoral district. Via primaries, nominations, donations, and other political mechanisms, the local opinions will be transmitted to the level of counties, from there to the state level, and then to the national campaign. At each step, an ever smaller set of issues has a chance of surviving and propagating. A small town may really care about the level of funding of its police force, but if that is an isolated concern, it will not go any further. For it to become a national issue, police funding at all levels of government must be at stake, in which case, the "renormalization" procedure will distill and amplify it to a major polarizing question of the election.

In 1980, the prestigious Wolf Prize in physics was awarded to the three discoverers of the idea of renormalization, Michael E. Fisher, Leo P. Kadanoff, and Kenneth G. Wilson. The latter was also awarded the Nobel Prize in physics in 1982. By then, renormalization was a go-to tool in physics, a sort of automated approach to find the degrees of freedom that really truly matter.

On a personal note, I have been hooked on the study of renormaliza-tion ever since being introduced to it by my then PhD advisor, Mikhail Lyubich in the early 1990s. Following physicists, mathematicians have dis-covered that applying this method to mathematical toy models of the phys-ical world works – and works surprisingly well. Similarly to physics, the renormalization approach has revolutionized much of mathematical mod-eling. Rigorous justifications – proofs – that renormalization distills the multitudes of degrees of freedom in renormalizable mathematical models to only a handful are difficult, to say the least. Some of the most beatiful and challenging math I know has been developed in pursuit of them (and I am happy to have contributed to its synthesis). Our understanding is in its very early stages. We cannot reliably point out when a model would not be amenable to renormalization, and thus to comprehension – as it would have too many, possibly infinitely many "important" degrees of freedom in it. In fact, in most of the successful mathematical investigations of renor-malization so far, things boil down to a single dimension – which is great if you would like to understand the model, but not so great if you would like to understand how to describe models which *one could not understand*. We are only at the beginning of the journey here, and do not yet have a math-ematical warning sign for scientists saying: "don't bother, you'll never be able to figure this out, the number of important degrees of freedom is too high". For now, at least, we can only speculate when this end of knowledge may have been reached.

Much shouting

> Where there is shouting, there is no true knowledge
>
> ---
>
> Leonardo da Vinci

By definition, any physics we can ever understand is renormalizable. Can there be aspects of nature that are not renormalizable and hence that we can never hope to understand? And I do not mean that a particu-lar mathematical model of reality is non-renormalizable and has infinitely many, or a finite, but a very large number, of significant degrees of freedom. Such unwieldy theoretical models have appeared in the past (think of the Ptolemaic orrery in Figure 17), but, assuming a more efficient way of describing reality exists, they are eventually discarded and forgotten. No, I mean the scary possibility that some natural phenomena may themselves be non-renormalizable, and thus not possible for us to comprehend with whatever theory we may formulate.

The long struggle to describe the Solar system illustrates what happens when science hits a wall. Clumsy empirical models with lots of moving parts are fitted to experimental results, only to discover that even more moving parts are needed to account for new experimental results – with no end in sight. Divorced from experiment, theories become objects of belief and are supported by theological arguments. *Ad hominem* attacks abound. One does not have to look far to see parallels in some of the struggles of modern physics. Particle physics, which was revolutionized in the early twentieth century by the invention of quantum mechanics, and has gone from one success to another in decades since (think nuclear bombs) seems to have hit a wall trying to understand the small-scale structure of the Universe. It currently relies on a grandly named "Standard Model" which, like a medieval orrery, has lots of moving parts, with more being added periodically. It lacks the framework which accounts for such fundamental phenomena as gravity. In a scientific vacuum, some of the theories develop based on their mathematical beauty and consistency, rather than any experimental value – String Theory is a prominent example. Scientific debates take a personal tone (for instance, Peter Woit's book on the shortcomings of String Theory is titled "Not even wrong..." [23]). Ultimately, these debates seem to be settled in the realm of academic politics.

Optimistically, these are healthy growth pains, and we are just waiting for the discovery of new laws of physics. But maybe, just maybe, we have reached the limits of human understanding? Albert Einstein's famous complaint about the appearance of statistical uncertainty in quantum physics was "God does not play dice with the universe". But what if He plays 170-dimensional chess instead, and we would need to reconstruct all of the 170 meaningful degrees of freedom to follow the moves? No one has promised to us, after all, that all of the mysteries of nature can be explained by a few mathematical formulas. It would be a true miracle if that is the case.

Chapter 3

The noise and the signal

> So easy to understand are all truths, once they are discovered; the point is in being able to discover them.
>
> Galileo Galilei (1632)

In spite of the doom and gloom of the previous section, we all know that science has been generally successful in reducing the mind-boggling complexity of the world surrounding us to concise mathematical formulas (or "laws of nature" as we usually call them). Yes, it is entirely possible that we may have hit a wall of non-renormalizability[1] here and there. It is also true that some mathematical models, while seemingly plausible, describe the world using too many degrees of freedom to be useful. And yet, simple and low-dimensional mathematical laws have been surprisingly effective to dive deep into the workings of the universe. This "unreasonable effectiveness of mathematics"[2] has driven many a scientist to philosophy or religion.

---

[1]Recall from the previous chapter that a natural phenomenon is renormalizable if it is possible to distill a few truly relevant parameters from the infinitely many variables potentially describing it. A non-renormalizable phenomenon could never be understood.

[2]This expression was coined by the American physicist Eugene Wigner in 1960 in his famous and provocatively titled article "The unreasonable effectiveness of mathematics in the natural sciences" [22] which captured some of the wonder of the... well, of the mathematics' unreasonable effectiveness, which we neither understand, nor deserve.

I would like to set this controversial, and by its nature speculative, topic aside. Instead, let me discuss the unreasonable effectiveness of *natural scientists*, armed with mathematics, to discern order in the overwhelmingly chaotic world and formulate simple laws which govern its function – at least when the laws are there to be found. In some sense, Kepler's finding of the mathematical rules governing the celestial clockwork, for instance, is no less amazing than the existence of Kepler's laws. We know intuitively that great scientists deserve to be put on pedestals, awarded Nobel prizes, and paid big bucks in cushy academic positions – because it must be hard to discover all of those wonderful laws of nature. And this is true even, and especially, if these laws are mathematically simple. Despite what lay people may think, the true difficulty in the natural sciences is not to develop a faithful description of an exceedingly complex natural world, but to *reduce* that complexity to exceedingly simple mathematics. Nothing else will do.

How difficult is it to find the needle of a mathematical law of nature in a haystack of data? Is there a systematic way of doing it? Can AI improve, or even replace, scientific discovery? To get some ideas, it will be good to start with a specific example.

The science of fitness

> ...it s important to ... not use
> astrology as a definitive tool to
> predict weight-related issues.
>
> "Astrotalk" astrology website

At the time of writing, the CDC[3] defines having a healthy weight as having a Body Mass Index (abbreviated by its first letters as BMI) between 15.5 and 24.9. Lamentably, the average BMI for adult American males is 26.6 and, for adult American females, it is a teeny bit lower at 26.5. Those numbers are in the "overweight" cathegory according to the CDC. The BMI is a remarkable measurement – it is the ratio of a person's body weight to the *square* of their height[4]:

$$\text{BMI} = W/H^2.$$

Its use suggests that for an adult the weight is supposed to *scale* as the square of the height:

$$W = \beta H^2.$$

Indeed, if this were the case, and everyone had the "ideal" weight for their height, then the ratio $\beta = W/H^2$ would be the same for every adult. The deviations from this hypothetical ideal can be measured by increases or decreases in β. This justifies the importance of the ratio β which is then labeled "BMI".

What makes this simple-looking measurement remarkable, you ask? If you double the size of a metal ball, then its mass will increase by a multiple of eight. The volume of the sphere of radius r is $\frac{4}{3}r^3$, so its weight increases as a *third* power of the radius. The same is true for a scaled copy of any familiar object. And yet, W scales as a square, not a cube, of H. Objects, whose areas, volumes, or weights scale as a "wrong" power of their sizes are known as *fractals*, a word which we associate with astonishing structural complexity. BMI implicitly hints at the fractal organization of a human body.

---

[3]Centers for Disease Control and Prevention, the US national public health agency.

[4]The weight is measured in kilos and the height in meters – the ranges of values would have been different if imperial measurements were used instead.

☛

This scaling rule was discovered (and BMI was defined) by a Belgian statistician Adolphe Jacques Quetelet in the first half of the 19th century by a careful analysis of multiple measurements of multiple humans. I do not know how he derived it, but it would have been natural to first assume that

$$W = \beta H^p$$

for $p \approx 3$. The non-linear formula can be turned into a linear one by applying the logarithm on both sides:

$$w = b + ph,$$

where $w = \log W$, $H = \log h$, and $b = \log \beta$. This can then be verified by measuring the heights and weights of a few hundred adults, and fitting a straight line to the resulting set of data points. In Statistics, this method is called *linear regression*, and it is both the simplest and the most popular tool of data analysis to this day. Regardless of what your initial guess for the value of p was, the slope of the best fitting line will be close to 2, not to 3. Have a look at the next figure, which shows how linear regression would work with a group of a hundred people.

☛

The quadratic scaling relation between height and weight is an obvious scientific success. It is a simple enough "law of nature" which is easy to verify using basic statistical techniques. It has important, although controversial, applications to human health sciences via the formula for the BMI. The discovery of this scaling law was based on a key hypothesis: *the body mass should scale as a power of the height*. This idea may initially be arrived at via a mistaken analogy between a human body and a solid metal object, a sphere, for instance, as I have described above. This particular parallel suggests a wrong scaling power, third instead of second. But no matter – as soon as you are on this track, and have access to the measurements of heights and weights of a largish group of people, the standard linear regression will set you right.

However, the power scaling hypothesis is by no means obvious. I am willing to bet that most people have never thought of the height-weight

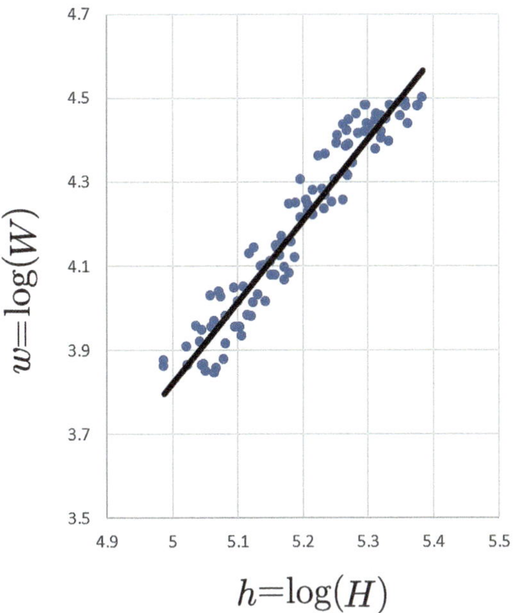

A simulation of the distribution of weights and heights of 100 people on a logarithmic scale. The linear trend is shown, its slope is very close to 2.

relationship in these terms, even if they know how to calculate the BMI. For a scientist to come up with it requires a leap of creative imagination, something that is not really taught in textbooks. But what if, instead of relying on such undefinable scientific creativity, we leaned on the data itself to extract this scaling law? It is, after all, so simple that any reasonably clever observer should be able to glean it through the forest of data. Right?

Fit or not?

> With four parameters I can fit an elephant, and with five I can make him wiggle his trunk.
>
> John von Neumann

I invite us to carry out a thought experiment. Let us imagine gathering some data on the weights and measurements of a group of a hundred or so people. A second-year Math class of an average size would do. We would record the students' weights, heights, and various other measurements (we do not know what the ultimate expression for the law would be, so it is prudent to collect a few of the parameters describing human bodies). Naturally, we would want to protect the students' privacy, simply to avoid possible embrassment when reporting something as personal as their body weights. So we will anonymize our list, by not using the students' full names. We could use their first initials, or, say, several digits at the end of their student numbers. Let us go with the three last digits of the student number as the more impersonal of the two, although the effect would have been the same with the initials. Once our list of data is ready, we will feed it to a data mining statistical software to discern the patterns predicting the ideal body weight.

What do you think the statistical analysis will focus on? Well, the second power of the height does strongly predict the weight. This prediction is not perfect, of course. The BMI varies even within the generally fit student population. In contrast, in our second-year Math class, the last three digits of the student number will be a nearly perfect predictor of the weight. The reason for this is silly and simple enough – they suffice to identify the students more or less uniquely (there may be some coincidences, of course, but they are not likely). So the software will have much more success predicting the weight with those, and hence will focus on them. And no other data will really be needed to predict the weight of the student, so they will likely be completely ignored. Even if we nudge the statistics in the right direction by only recording two columns of the information: the three identifying digits and the height, it will probably focus on the former at the expense of the latter. After all, why trade a perfect predictor for an imperfect one?

By using the three digits of the student number, the data-mining algorithm will be able to come up with an excellent rule for predicting the students' body weights. It will match their weights exactly. It will also be laughably useless for the general population.

You could correctly point out that there is no reason to assume that the student numbers would have anything to do with body weight. But that is the whole point – we were not supposed to know in advance what did and what did not, the data were supposed to speak for themselves. And even if we had excluded such descriptors, there would have been enough other data points in our list: birth dates, Zodiac signs, colors of the eyes, and so on to identify the students uniquely. Any such combination would work as a model to predict the body weight in our sample group. The model would probably be complicated, but it would work perfectly well – that is, until we would try to apply it to the human population at large.

In the scientific literature this unfortunate situation is known as *overfitting*. Let us unpack what is implicitly suggested by this term. There is a "true" law of nature, that says that the ideal body weight is proportional to the second power of the height. This is a renormalizable law: all of the innumerable other parameters describing the students in our group are pretty much irrelevant to the description. The statistical signal sent by the height is all-important, everything else is, at best, somewhat helpful, and at worst useless or misleading, such as the students' Zodiac signs. However, and for obvious reasons, the BMI ratio will not be constant in our group, which means that a single quadratic formula $W = \beta H^2$ is not going to be a perfect match, no matter what value β we use. On the other hand, there will be combinations of descriptors which will give better, maybe even exact matches. They may not even include heights at all, and consist, for instance, of the combinations of the Zodiac signs, eye colors, and lengths of the pinkie fingers. And they would produce seemingly better "laws" of nature, which would match well to the data in our group. But, of course, these laws would be generally useless outside of our group. Their better **fit** to the data would be accomplished by adding generally irrelevant parameters which would describe this particular small data set **overly** well, and hence would present themselves as the correct solution to a naïve data miner.

The difficulty should be clear by now – even if a simple mathematical law of nature exists, the evidence for it is drowning in the sea of data. Recognizing what is relevant and what is not requires a conceptual scientific insight. Generating such an insight is what scientists are paid the big bucks for. And where they very frequently make conceptual mistakes, which result in seemingly good and seemingly sensible models, which seem to agree with the existing data – and yet are profoundly wrong. In fact, as we have seen above, a sufficiently large number of sufficiently diverse measurements is pretty much **guaranteed** to work in fitting the data set, no matter how irrelevant these measurements are to the actual phenomenon they purport to describe. One of the most cited papers in the history of modern science is the article by a Stanford biostatistician John Ioannidis titled "Why Most Published Research Findings Are False" [7]. In it, he attempts to summarize how a wrong conceptual insight may arise in science (and regularly does), and how easily it is then validated because of seemingly matching the data.

Kolmogorov's razor

> If people do not believe that
> mathematics is simple, it is only
> because they do not realize how
> complicated life is.
>
> John von Neumann

The correct conceptual insight into what parameters are relevant is thus key to producing sensible mathematical descriptions – laws – of nature. Everything else, but this insight, can be easily automated. **Can the insight itself be automated?** The process of renormalization is supposed to distill the relevant parameters and serve as an automation tool. This is an important thought, and we will get back to it in a bit. But first, what do we even mean by a sensible law of nature? How do we know that we have found the right set of descriptors, which do not suffer from overfitting?

One of the greatest mathematical minds of the 20th century, A. Kolmogorov, who, in particular, was one of the founders of modern Probability Theory and statistical science, was deeply concerned with this question. In a visionary insight at the very end of his scientific career, Kolmogorov proposed judging the models by a version of what is known as *Kolmogorov*

complexity. To explain this fundamental mathematical concept, let me start with an example.

Look at the real number x below:

$$x = 0.12345678910111213141516...$$

Can you guess what the next two decimals of x should be? It looks like it would be 1 and 7, following the simple rule for generating the digits of x after the dot:

```
write out all natural numbers in a row
```

Kolmogorov's blindingly simple insight was that one can use *the length of the description* as a measure of the complexity of any object: a number, an image, a written text, etc. From this point of view, the number x is as simple as it gets, since all of its infinitely many digits can be generated by a finite instruction. That instruction carries all of the information you need to write out x, and it is short – finite – infinitely shorter than x itself.

Using modern terminology, we were able to *compress* x to a finite instruction. Most people are familiar with the idea of compression from using apps, such as ZIP, to "make a file smaller". The feats these apps perform consist in finding things like repeated patterns inside the file which can be replaced by their much shorter descriptions. Kolmogorov complexity quantifies how much the size of a file can be compressed. The following simple exercise should convince you that *compression is not always possible* – there are objects which do not allow any simple descriptions.

Let us switch to binary for a moment, since that is how computers think, and imagine a string consisting of n binary digits, each of which can be either 0 or 1. Since there are two choices for each of the n digits, there are exactly 2^n possible strings we can write in such a way. Now, suppose we would like to make a string shorter, by compressing it to a smaller description. The description would again be written in a binary code, and to be shorter, it would have to contain at most $n - 1$ binary digits. How many such descriptions are there? There is exactly one with no digits at all, two with one digit ("0" or "1"), four with two digits ("00", "01", "10", "11"), and so on. There are 2^{n-1} description strings with $n-1$ digits. A compression which makes the initial string shorter by a single digit is probably not worth the trouble, but we will allow it for completeness. Thus, the number of all possible descriptions is at most

$$1 + 2 + 2^2 + \cdots + 2^{n-1} = 2^n - 1,$$

one less than the number of all length-n strings[5]. Hence, at least one string must have the maximal Kolmogorov complexity – it cannot be compressed at all.

Reducing a natural phenomenon to a simple physical law is, in fact, a problem of compression. We can make a list of all of the relevant measurement data that we have collected, put it into one huge list, and then ask if there is a *simple* rule generating this list. There is a complication, however – we cannot hope for a truly perfect match of the law to the experiment because our measurements will always contain some random noise coming from errors of measurement. So we should aim for a compression rule that will explain most of the data, except for the noise. The difficulty this poses should be evident: how can we tell whether the mismatch to the data is truly random, and does not come from our description being sub-optimal? Perhaps a better compression technique would be available which will explain away much of the "noise"?

Kolmogorov comes to the rescue here as well, with the idea of *Kolmogorov randomness*. In a nutshell, a truly random sequence of events cannot have a simple compression rule. The sequence of all natural numbers is not random – and the number x above cannot be generated "randomly" in the Kolmogorov sense.

There are plenty of computer algorithms, some of them quite short, to generate seemingly random sequences of numbers. For instance, the famous American mathematician and pioneer of computing John von Neumann, in 1946, while working on the Manhattan project, suggested the

---

[5]If we let $S_n = 1 + 2 + 2^2 + \cdots + 2^{n-1}$, then

$$2S_n = 2 + 2^2 + \cdots + 2^{n-1} + 2^n = S^n + 2^n - 1.$$

Subtracting S_n on the left and on the right, we get the formula $S_n = 2^n - 1$.

following approach, known as the middle square method. Take a natural number with, say, eight digits. View the number consisting of the middle four digits as the seed of your "random" sequence. Square it – and if the result has less than eight digits[6] add zeros in the front to bring the number of digits back to eight. Now take the middle four digits as the next number in the sequence, and use it to repeat the calculation. And so forth. For some initial seeds, although not for all of them, this works spectacularly well, producing a very long sequence of seemingly random numbers. Von Neumann found the results sufficient for whatever (highly classified) needs he had. But from Kolmogorov's point of view such a sequence should not be viewed as random, since it can be described by a simple rule[7].

Armed with Kolmogorov's ideas we can now say how to measure the optimality of a natural law – that is, of the mathematical model explaining the experimental data. Namely, the optimal model should have a short description – the shorter the better, which can be measured by the length of the computer code which goes into a) encoding the model, and b) describing the data set to the model. And, crucially, the deviation of the model's predictions from measurements should be caused by truly random fluctuations. These fluctuations cannot be produced by a pseudo-random number generator (or, at the very least, cannot be produced by a pseudo-random number generator which also has a short description). This last requirement ensures that we could not produce a better fit to the data by using a different description, which is similarly short.

An optimal description in the sense of Kolmogorov should be seen as the "best" among all the possibilities. We could call this *Kolmogorov's razor*, as a precise mathematical version of Occam's razor. This famous philosophical principle was named after William of Occam, the 14th century English philosopher, who was fond of citing it in his work. Quoted

---

[6] It cannot have more – can you see why?

[7] Somewhat pedantically, computer scientists have taken to calling such programs *pseudo-random number generators* to emphasize that no sequence generated by a fixed rule is truly random.

in Latin as *Entia non sunt multiplicanda praeter necessitatem*, it literally translates into "Entities must not be multiplied beyond necessity", taken to mean "The simplest explanation is usually the best one".

The rule $W = \beta H^2$ is an excellent example. Taking this point of view, we would have to accept as a true law anything that passes Kolmogorov's challenge, without any prejudice. So if the Zodiac signs could be used to give an equally simple description of W, then we would have to seriously reconsider the prevailing scientific attitude towards astrology (jokes aside (?) maybe the Zodiac signs are relevant here, at least according to a venerable American magazine, as seen below).

A headline from the online edition of *The Reader's Digest*.

This notion of the complexity of a model has been made formal in a relatively recent specialist paper [20]. Sadly, in the same paper, the authors proved that the general problem of finding Kolmogorov's best model from a set of data is algorithmically unsolvable. That is, a computer program which would automate finding such models cannot exist. That may be a relief for some scientists, who can continue publishing mostly wrong papers with the guesses on how the universe is run. But it is hardly a solace for the rest of humanity. So what is there to do?

Let us make another thought experiment. Take the noisy data from figure 19, representing the dependence of human weight on height in the logarithmic scale. It does not quite fit on a straight line, but bounces around it randomly. Let us *average* the nearby data points. We could, for instance, take the horizontal range of the h variable, divide it into twenty or so intervals of length of about 0.02, and then average the h and w values in each. Or we could group the points on the plot into clusters of, say, four nearest neighbors and average those. Or we could do something more creative, like spiltting the picture down the middle, moving two halves together, averaging the points, and then stretching the graph back to full size. Since the noise around the trend line is random, its overall average is zero. No matter which of these averaging techniques we apply, the result should be the same – less noise, more signal, a better fit to the line.

Now imagine doing the same with student numbers, Zodiac signs, or any other spurious data we may have collected. Since there is no signal to be found in their relation to the body weight, averaging is going to destroy any appearance of a trend. It will result in a horizontal line – a dead signal.

This is, of course, the essence of renormalization. It gives hope that a renormalizable law of nature can be found in the haystack of data by successive filtering of the spurious data by averaging. In the next chapter, I will try to convince you that this is exactly the process by which a mind – natural or artificial – makes sense of the Universe.

Chapter 4

Live and learn

The pleasures arising from
thinking and learning will make
us think and learn all the more

Aristotle

We can't learn without pain

Also Aristotle, likely in a bad
mood

I have worked hard to convince you that the world is complex, mysterious, and difficult to describe. It is surprising that humans have managed to understand much about its workings. It is even more surprising that they function pretty well in it even without fully understanding its workings. Walking upright and not falling down has somehow been manageable before Newton's discoveries of the laws of motion and of gravity. In fact, the attempts to design walking robots based on direct applications of these laws have mostly failed: walking without falling is a complex endeavor, it involves interactions of too many moving parts.

Trying to take into account the mechanical positions of robotic limbs, the angles at which they bend, the gait, etc, etc, all based on the laws of Newtonian physics involves real-time adjustment of too many parameters. As an apt analogy, think of the skill required of a puppeteer to make a wooden puppet come to life. Imagine, how hard it might be to make the puppet appear to walk "normally" across the room by pulling the strings attached to its limbs – the choreography would be overwhelmingly complex. Recent successes in making robots walk upright are based on an entirely different approach – *teaching* them, as we would teach a human baby.

Somehow, our brains are capable of *learning* – that is, discovering "laws" of the world from empirical observations. Something that science and scientists often struggle to do on a conscious level.

Imagine looking about an unfamiliar room and forming a plan for crossing it from one corner to another without bumping into furniture or stepping into the bear trap someone has set in the middle of the carpet. It is an extraordinarily complex task. One has to have a general idea of a "room", and "furniture", and even a "bear trap". Then one needs to parse the visual data capturing literally everything in the room into a map. Armed with this knowledge, and with a mental model of "walking" in mind (putting one foot in front of the other, keeping ones balance, etc) one can then march across the said room. The difficulty of distilling infinite amounts of data that our sensory networks can absorb into a coherent, low-dimensional set of "laws" of walking across the room is what makes learning hard. Yet we do this seemingly without thinking. A dog could learn to do it without any difficulty, assuming, of course, the dog has been previously taught that it is not a good idea to step into a bear trap.

Making robots capable of learning has strangely turned out to be an easier task than manipulating them as puppets. The story of Pinocchio is often dredged for deep psychological metaphors. But perhaps the true metaphor of the story lies on the surface – the wooden puppet had to go into the world and learn from his own mistakes to "become a boy", like all of us do.

I want to talk about the magical mystery of learning. Partly to convince you that the unconscious learning that a brain does has much in common with what we think of as "scientific process", and is subject to the same mathematical constraints. Partly to explore how a machine can learn, and whether AI is subject to the same limitations as we are. But I will start with a seemingly unrelated, although fun and fascinating mathematical topic. Trust me, it will be relevant.

Fractals

Intuitively, the word "fractal" describes an object which has intricate details at every scale. By way of example, popularized by the late Benoit Mandelbrot, the coastline of Norway is nearly impossible to map, so carved it is by fjords. And increasing the resolution of a map does not help, as the coastlines of fjords are similarly notched by smaller fjords, whose coastlines, in turn, consists of even smaller fjords. In a now famous quote, Mandelbrot noted that "Clouds are not spheres, mountains are not cones, coastlines are not circles, and bark is not smooth, nor does lightning travel in a straight line." All of these are fractal shapes, revealing undiminished complexity under magnification.

Setting poetry aside, let us look at a specific, mathematical example of a fractal shape. Meet *the Barnsley fern*:

It was presented by a British mathematician Michael Barnsley in his beautifully printed 1988 book "Fractals Everywhere" [1]. In the spirit of the above quote of Mandelbrot, it illustrated the fractality of a fern leaf by producing its mathematical imitation. It is a true fractal. Have a look at the next figure, which shows four zoom-ins into the image of the fern. I made them in a hurry, and honestly do not remember what the magnification factor was for each of them. But it would not matter, as the image does not get any simpler (or harder) under magnification. As mathematicians say, it is *self-similar*. A Barnsley fern appears to be an object of an exquisite complexity, conveying the idea that Nature itself is infinitely complex to describe, as it is filled with similar fractal shapes (clouds, mountains, coastlines, bark of trees, and, well, ferns). And yet, in the next paragraph, I will try convince you that the opposite is true. *The Barnsley fern is spectacularly simple to describe.*

Several arbitrarily chosen magnifications of the Barnsley fern.

Don't believe me? Well, here is its description:

$$F_1\begin{pmatrix} x \\ y \end{pmatrix} = \begin{pmatrix} 0 \\ 0.16y \end{pmatrix} \qquad F_2\begin{pmatrix} x \\ y \end{pmatrix} = \begin{pmatrix} 0.85x + 0.04y \\ -0.04x + 0.85y + 1.6 \end{pmatrix}$$

$$F_3\begin{pmatrix} x \\ y \end{pmatrix} = \begin{pmatrix} 0.2x - 0.26y \\ 0.23x + 0.22y + 1.6 \end{pmatrix} \qquad F_4\begin{pmatrix} x \\ y \end{pmatrix} = \begin{pmatrix} -0.15x + 0.28y \\ 0.26x + 0.24y + 0.44 \end{pmatrix}$$

I have to explain myself. As it turns out, the four two-dimensional linear functions F_1, F_2, F_3, F_4 encode the self-similarities of the Barnsley fern. Each one of them is responsible for taking a part of the fractal image and magnifying it to match the whole. Knowing these self-similarities is knowing the Barnsley fern – no other fractal shape corresponds to same exact magnifications.

Armed with these formulas, it is easy to compute an image of the fern. One can start with an arbitrarily chosen dot in the plane, and then start applying the four rescalings to it again and again, every time selecting which one out of the four to apply at random. This will produce a sequence of dots, which, after enough of them are generated, will look like the Barnsley fern; the more dots, the sharper the image.

Recall Kolmogorov's Razor: despite its apparent visual complexity, *the Barnsley fern is simple because it has a simple description.*

A teachable moment about learning

> Out of clutter find simplicity
>
> Albert Einstein

The process by which our brains develop an understanding of the world is what we know as *learning*. Think of learning to recognize images: cats, dogs, people, or ferns. There is an infinite variety of visual shapes that these objects may present. Our brains cannot possibly contain an exhaustive library of all of the possibilities, and try to match what we see to it. We would both run out of memory capacity and of processing time. The above discussion of the Barnsley fern gives an excellent toy model of learning. The fractal complexity is reduced to a simple set of rules – its four self-similarities. The process of applying successive magnifications until the fractal image comes into focus is an instance of *renormalization*. Yes, you have read it right – the same principle applies here as to the laws of physics. For an image to be recognizable, it must be renormalizable – that is, all of its intricate details, variations, perspective, lighting, angle of view, and so on, must be reducible to a few basic features. The process of distilling these basic features from the noisy, messy visual data is learning.

I realize that this may sound like a bold, overly general, and perhaps scientifically controversial claim – how could I possibly be sure that this is how the brain *actually* learns? In my defense, I am going to present some pretty convincing evidence.

But before I embark on this quest, let me point out that these basic principles of reducing visual complexity are commonly put to use in *image compression*. An extreme example is so-called *fractal compression*. As the name suggests, it aims to reduce an image's complexity by representing it as a fractal. A fractal can be conveniently encoded by its self-smilarities, allowing for very efficient compression of information. Thus, the image of an actual fern can be replaced by a version of the Barnsley fern, with the coefficients of the functions F_1 through F_4 adjusted to match. A real fern leaf is not a true fractal, and if we zoom sufficiently into its edge, the self-similarity will disappear. But, looked at on a large scale, it can be beautifully represented by a fractal shape and thus by a very simple description. Encoding the image by different descriptions on a decreasing sequence of scales brings us into the realm of modern image compression tools.

A similar principle is followed by any compression algorithm, not only for images, but for any data. Some data is truly "incompressible" – it is maximally complex in Kolmogorov's sense, and cannot be reduced or renormalized. It is the enemy of learning, just as a non-renormalizable physical theory would be. It is simply impossible to reduce such data to a manageable size, so it can only be memorized.

Here is some scary food for thought: *a truly random data is incompressible.* If we had a way, for instance, to randomly stick a needle into the number line, and write out the digits of the number it hit, then there would be no pattern to them at all.[1]

True randomness is a nemesis of learning. The fact that we can function in the world and be taught to comprehend it both at a conscious and an unconscious level, suggests an extreme lack of randomness – *nothing is purely by chance.*

---

[1]A precise way of making a probabilistic statement like this would require adding "with probability 1" at the end of the sentence. But, of course, in real life we think that an event which has probability 1 will happen *for sure*. Its opposite, which has probability 0, has a zero chance of ever being observed.

Presenting the evidence

> Some circumstantial evidence is
> very strong, as when you find a
> trout in the milk
>
> Henry David Thoreau

> People almost invariably arrive
> at their beliefs not on the basis
> of proof but on the basis of what
> they find attractive
>
> Blaise Pascal

One of mathematicians' favorite ways to characterize fractal shapes is by their *dimension*. Here is a recipe for measuring the dimension of a given image. Superimpose over it a regular grid consisting of squares of some fixed size ϵ. Then count how many of the grid squares intersect with it; call that number $N(\epsilon)$. For a true self-similar fractal, it turns out that $N(\epsilon)$ is proportional to a *power* of $1/\epsilon$:

$$(4.1) \qquad\qquad N(\epsilon) \approx C \cdot \left(\frac{1}{\epsilon}\right)^{d} = C \cdot \epsilon^{-d},$$

where C is some constant that does not depend on ϵ. The approximate equality means that the ratio of the logarithms of the two sides converges to one, as ϵ gets smaller.

Or, if the formula does not speak to you: if you reduce the size of a grid square M times, then the number of squares intersecting with the image will go up by, roughly, M^{d} times.

For "smooth" shapes in the plane, the constant d would be equal either to 1 (if they are curves) or to 2 (if they are filled-in regions). For a fractal, d is not an integer. It is fractional, and that is the origin of the word "fractal".

I must explain the reason why the number $N(\epsilon)$ *scales* as a negative power of ϵ. One of the ways of expressing the self-similarity of a fractal is that it "looks the same" under any magnification. I honestly forgot what magnifications I used to produce the zoom-ins into the Barnsley fern in the

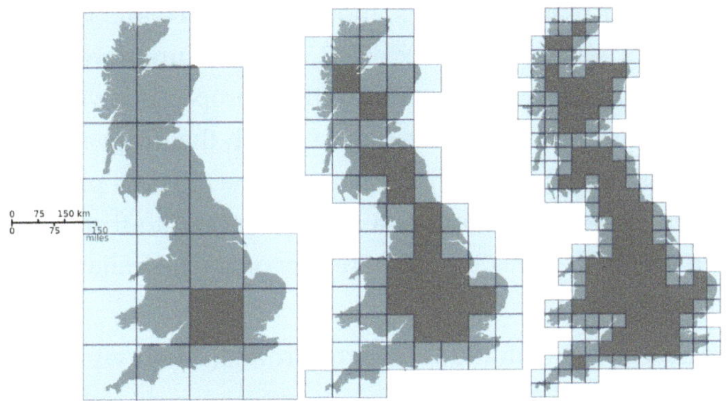

I could not resist stealing from Wikipedia this great illustration of measuring the fractal dimension of the coastline of Britain (image by Prokofiev, Wikimedia Commons)

figure on page 55, and now it is too late – there is no way to recover that information from the images themselves. Fractals are *scale-free*, that is, one cannot recognize the scale at which one is looking at them. This means, of course, that any conceivable measurement of a fractal must be scale free as well, otherwise it could be used to identify the degree of magnification. So the graph of $N(\epsilon)$ should look the same under linear rescaling of the horizontal and vertical axes, which translates mathematically into

$$\alpha \cdot N(\epsilon) = N(\beta \cdot \epsilon)$$

for some positive values of α and β, where $\beta > 1$ is the magnification factor. As it turns out, the only such functions[2] are power functions. Scale-freeness forces a power-scaling law.

Power-scaling laws are the best indirect evidence of self-similarity; when observed in Nature, they strongly indicate "fractality". As I posited that brains learn by reducing the world to self-similar structures, it would be natural to ask whether brain function exhibits power-scaling. And power-scaling laws abound in measured brain function, leading researchers to

---

[2]The only "nice" such functions – that is, analytic ones. It is possible to produce pathological examples of functions which satisfy the same equation, but, while mathematically interesting, they are not something one would expect to encounter in Nature

conclude that our brain activity is, indeed, scale-free [6]. For instance, as the brain is processing a complex event, firing neurons would not fill its volume uniformly. Roughly speaking, their proportion would vary as a power law with the size of the region in which the firing activity is measured. If we think of the neuron activation as the physical process of learning – burning an interpretation of a phenomenon into the brain – then it very much seems that this interpretation is of a self-similar nature, like a fractal encoding.

This perfectly fits with the idea that learning (conscious or not) is, at its core, renormalization.

We cannot truly look "under the hood" of a functioning human brain to confirm this. But we have no such constraints with artifical brains. The current explosive success of artifical intelligence designs is based on a revolutionary concept of *deep learning* – and it is here, where the strongest direct evidence comes from.

Playing chess against an alarm clock

> A Little Something to Offend
> Everyone: History of the World
> Part 1
>
> ---
> The title of a 1981 film by Mel
> Brooks

Before a deep dive into deep learning and artificial intelligence in general, it would be useful to look at an undisputable high point of the development of non-intelligent computers. This great computational feat happened in the late 1990s – an IBM produced computer *Deep Blue* beat the then world chess champion Garry Kasparov in a match of six games. Chess is an incredibly complex game, and chess champions have deservedly been considered the most rare of prodigies. And yet, Kasparov disparaged his computer opponent for being "as intelligent as your alarm clock". This sounds pretty harsh, but a case can be made that it was accurate.

I am not going to even try defining what intelligence means – there are many thousands of pages of scientific articles written by very intelligent scientists giving all sorts of definitions. I am just going to point out that chess is a particularly challenging game to program a computer for. There is no grand "strategy" to win in chess, no guidelines to follow to stay unbeatable. We can say that chess is not a renormalizable game – it cannot be distilled to a concise set of principles. That is what makes it so fun and special, of course. But it means that there is no "learning" chess[3]. The process of becoming a chess master is based on painstaking practice. To a large extent, it is brute force memorization – either of entire games, or of general somewhat heuristical "strategies", as well as brute force calculation of the tree of future possible moves (only a few moves ahead, as there are generally too many possibilities to consider).

Deep Blue was indeed made to memorize games, as well as heuristic strategies of several chess "coaches". This resulted in a machine of a staggering expense and complexity. In a sense, the complexity of the computer *mirrored* the complexity of the game. When it won, it did not mean that computers became better at chess than humans. That particular computer beat that particular chess champion in that particular match. It did not learn to beat all future humans, since it did not learn, period. If we think that intelligence corresponds to ability to learn[4], we may see some truth in Kasparov's "alarm clock" remark. In *Deep Blue*'s defense, if it was a clock, then it was an exquisitely complex one – *replicating* to some extent the exquisitely complex clockwork of playing chess.

Pardon me for this little interlude, but it does give some idea of the challenges of creating an intelligent machine. It is the same type of challenge that human understanding faces. The universe is infinitely complex. Memorizing this complexity would require an infinitely complex computer.

---

[3]As opposed to learning to play chess – which means memorizing a finite set of rules of the game.

[4]As well we should: in Computer Science, the subject of Artificial Intelligence became synonymous with Machine Learning.

The explosive progress of the AI revolution we are currently witnessing is based on an entirely different approach. Known as *deep learning* it has been scarily successful. And its workings are the clearest evidence that learning works by distilling simplicity from clutter.

Deep learning

> "O Deep Thought computer,"
> he said, "the task we have
> designed you to perform is this.
> We want you to tell us...." he
> paused, "The Answer."
>
> Douglas Adams

Surprisingly for the uninitiated, the adjective "deep" in "deep learning" does not stand for "profound". Rather, it is "deep" as in how a lake can be deep, or, rather, how a layered cake with many layers can be deep. It describes the structure of the neural network, which consists of several layers of artificial neurons – whose depth is the number of layers. Each layer is tasked with extracting some features, perhaps rather crude ones, and then passing them onto the next layer. This process progressively refines the knowledge, until some structure emerges from the data, which can be "learned".

A common example is automated face recognition. A high resolution grayscale picture of a face captured, for instance, by a surveillance camera contains hundreds of thousands, if not millions, of pixels. The number of ways of naïvely encoding a face would translate into the number of ways pixels could be shaded on a screen, which is a staggering number of combinations.[5] Not only that, but the same person's face would look different turned at an angle or in low lighting conditions, and we would need to teach a neural network to somehow identify combinations of pixels which could be turned one into the other by changing the point of view, etc. Treating images of faces as raw data is clearly not a way to go.

A layered or *hierarchical* approach – which is how our own visual cortex would handle the problem – would go something like this. Instead of trying to process the image of a human face as a whole, the first layer of neurons would just attempt to find the *edges* in the image – the boundaries between lighter and darker regions. This is a fairly straiightforward task, and building and training a set of neurons to do this does not require much

---

[5]Consider a 25x25 pixel screen with only two colors, black and white. The number of different images which can be represented with such ridiculously low resolution is greater than the total number of atoms on Earth.

effort. It can then pass this knowledge to the next layer, which will be trained to find recognizable features of faces made of these edges, such as eyes, ears, chins, and hairlines. To the artificial brain we are building, these will look somewhat cartoonish, since they will be made by assembling the edges found by the first layer, but that only makes the task simpler. Now, the following layer can take these features and combine them into sketches of actual faces. These faces can be identified by the positions of eyes, noses, ears, etc., and compared in the same way. Perhaps, the last step could be even optimized a bit: instead of looking at the positions of the features individually, the neural network might focus on some combinations (something like the square of the length of the nose minus the cube of the distance between the eyes) if they are more effective.

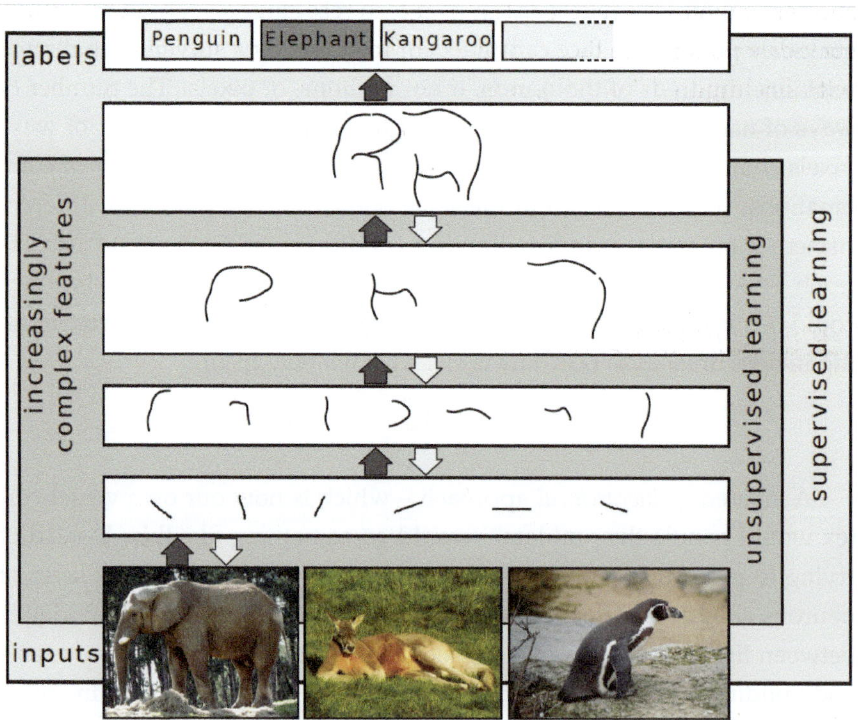

An illustration of the hierachical approach to image recognition in deep neural networks. Image by S. Behnke [17], reproduced with permission.

This hierarchical approach to feature recognition can sometime cause interesting optical illusions. No matter how long we stare at the checkerboard in the figure below, it would seem to us that square A is much darker than square B. This cannot be helped. By design, the first layer of our visual neurons identifies the edges between light and dark features. A is a darker feature surrounded by lighter features, and B is the opposite. Further processing by our brains finds a matching image – a checkerboard – at which point we *know* that A is darker than B. Fin.

The checker shadow illusion published by Edward H. Adelson, professor of vision science at MIT in 1995. Square A is actually the same brighness as square B. Source: Wikimedia commons.

By now, you may have noticed the parallels between our discussion of renormalization and the hierarchical, or "deep", approach to machine learning, which is responsible for the ongoing AI revolution. This has not gone unnoticed in the scientific literature either. In a 2014 paper [12], mathematical physicists Pankaj Mehta and David Schwab showed how to turn a particular kind of a deep learning neural network into a renormalization "machine". Many other examples of the connection between the two concepts soon followed.

Just think: cognition serves to interpret the world around us, which is a confusing mix of noisy data. To "learn" is to be able to reduce the infinitely complex sea of data to a few meaningful features. As long as the phenomenon we are attempting to learn is renormalizable, this is doable by repeatedly refining the data, amplifying what is important. Deep learning is indeed renormalization by another name.

Learning of any sort – conscious, unconscious, scientific, or even machine learning – follows the same path in distilling laws from the infinite complexity of the world. It therefore faces the same limitations. To be "learnable" in principle, a phenomenon must be reducible to a finite number of degrees of freedom. And that number better be very small, so an effective mental map of it can be designed. This is a fundamental barrier and, once reached, cannot be defeated by either human or artificial brains. AI will, in all likelyhood, be a great boon to civilization, but it will not help overcome this limitation.

And I cannot resist emphasizing once again what a miracle it is that we are able to learn so much about the world. A baby learning to walk and a physicist studying gravity would have been equally stumped if said gravity depended in a meaningful way on a handful more parameters than it does according to Newton's concise law of attraction.

Chapter 5

Out of order, chaos

> Chaos was the law of nature;
> Order was the dream of man.
>
> Henry Adams, an American
> historian

You may have been wondering when I would bring up the most telegenic obstacle to mathematical understanding, popularly known as Chaos. The point I have been hammering on in the previous sections was that the universe is hard to describe; its workings may be too complex to discern meaningful patterns in them. Chaos, on the other hand, refers to the paradoxical scenario when a simple mathematical law of nature may lead to unpredictable outcomes. In this way, it seemingly negates the success of the scientific endeavor, which aims to reduce natural phenomena to concise physical descriptions. What good would such a description be if it cannot be used to make any meaningful predictions or conclusions?

I would like to start talking about Chaos with a particular example, which is responsible for bringing it into the popular consciousness. It is familiar to all of us from our daily lives.

M. Yampolsky, *The End(s) of Knowledge*, Copernicus Books,
https://doi.org/10.1007/978-3-032-01423-8_5

A butterfly flaps its wings

> It's difficult to make predictions,
> especially about the future
>
> ---
>
> An old Danish proverb

> Perhaps Rain, Perhaps Not
>
> ---
>
> Josh Billings, 19th century
> American humorist, on weather
> forecasting

If you want to know whether the weekend will be rainy or not, you will find your local weather forecast somewhat evasive on the issue. Instead of telling you "yes, it will rain the coming Saturday" or "no rain", it will frame the answer in terms of likelyhoods and maybes. Have you ever wondered why the answer to a yes/no question is given in such an uncertain manner? Weather forecasts are produced by fairly sophisticated weather models which are run on fairly powerful computers. If a meteorologist inputs the description of today's weather conditions into the model, then the computation will give an unequivocal yes or no answer to the question whether Saturday will be rainy or not. So why are you not given such a forecast?

The answer is discouragingly easy: you are not being told what the weather model has predicted because this prediction is completely unreliable. Yes, you read this right – it is unreliable, almost to the point of being meaningless. And weather people have known it since the early 1960s. In 1961, an American meteorologist Edward Lorenz was using a computer to produce a weather forecast using parameters like temperature and wind speed. Lorenz was a professor at the Meteorology Department of MIT, and an expert in weather simulations. The computers of the day were slow. Lorenz wanted to extend a forecast he previously computed a bit further into the future. Instead of repeating the whole computation from scratch, letting it run for a longer time, he put in some intermediate values of the weather parameters that his program had previously output. For simplicity, imagine that he initially ran a 5-day forecast, and wanted to re-run it to get a 7-day forecast, but starting with the intermediate predicted weather for day 3. After all, that part of the work was already done – so it would be completely fine not to repeat it and just use the numbers that the computer had recorded.

To Lorenz' enormous surprise, starting with the values output for day 3, he would get completely different predictions for days 4 and 5, than he did with the initial run of his computer program. To Lorenz' enormous credit, he figured out the culprit. When he entered the intermediate data, for one of the variables he used the value 0.506 from the computer print-out instead of entering the full precision 0.506127 value that the program was actually using. The round-off error in the printout was tiny – about 0.025% – but it was sufficient to throw the calculation completely off and produce very different weather predictions.

In his book "The Essence of Chaos" [10] published in 1993, Lorenz himself described the discovery he had made as follows[1]:

At one point I decided to repeat some of the computations in order to examine what was happening in greater detail. I stopped the computer, typed in a line of numbers that it had printed out a while earlier, and set it running again. I went down the hall for a cup of coffee and returned after about an hour, during which time the computer had simulated about two months of weather. The numbers being printed were nothing like the old ones. I immediately suspected a weak vacuum tube or some other computer trouble, which was not uncommon, but before calling for service I decided to see just where the mistake had occurred, knowing that this could speed up the servicing process. Instead of a sudden break, I found that the new values at first repeated the old ones, but soon afterward differed by one and then several units in the last [decimal] place, and then began to differ in the next to the last place and then in the place before that. In fact, the differences more or less steadily doubled in size every four days or so, until all resemblance with the original output disappeared somewhere in the second month. This was enough to tell me what had happened: the numbers that I had typed in were not the exact original numbers, but were

---

[1]You will notice that Lorenz talks about days and months rather than hours and days as the timespan of his model. Such was the over-optimistic spirit with which weather forecasting was approached at the time.

the rounded-off values that had appeared in the original printout. The initial round-off errors were the culprits; they were steadily amplifying until they dominated the solution.

Lorenz's talk at the 139th meeting of the American Association for the Advancement of Science in 1972 was titled dramatically: "Does the flap of a butterfly's wings in Brazil set off a tornado in Texas?". Since then, such extreme instability of weather models became popularly known as "The butterfly effect". No measurement is carried out with absolute precision in Nature, and, in any case, a weather model is by necessity a simplified version of reality – so if an extreme change in predictions can result from a 0.025% change in its inputs, then predictions become well and truly useless.

This type of unpredictability has captured popular imagination under the name "Chaos". The dictionary definition of chaos is "complete disorder and confusion", which is hardly an encouraging thought. Another insight of Lorenz's was that chaos could appear in rather simple models. A famous example which he described in his 1963 paper [9], now known as *the Lorenz*

system, looks like this:

$$\frac{dx}{dt} = 10(y - x)$$

$$\frac{dy}{dt} = x(28 - z) - y$$

$$\frac{dz}{dt} = xy - \frac{8}{3}z$$

I do not mean to intimidate you with equations, but I would like you to see, without going into any details, how simple the math looks. The equations represent a model of atmospheric convection, in which t is the time, $x(t)$ is the rate of convection, and $y(t)$ and $z(t)$ are the horizontal and vertical temperature variations. Despite the simplicity of the formulas and the fact that there are only three variables, the model is chaotic. A tiny change in the initial values of x, y, z dramatically changes the model's predictions. Complete disorder and confusion indeed.

Butterflies make another appearance in this story. Suppose I would like to explore numerically the model described by the Lorenz system. Let me select a starting set of values $x(0)$, $y(0)$, $z(0)$ in the Lorenz system (the zeros in the parentheses indicate the starting time, $t = 0$). A computer simulation will then produce approximate values of x, y, z for any future time $t > 0$. A way of visualizing how these values evolve with time is to put them all on a three-dimensional plot[2]. It turns out that, for pretty much any set of starting values, for large t (that is, far in the future) the dots on the plot carve out the same shape. Uncannily, this shape resembles a butterfly, and has become known as *the Lorenz butterfly attractor*.

Thus, paradoxically, even though the values of x, y, and z at large time t would change drastically if the starting values change even a slightest bit, overall they trace out the same shape. The Lorenz butterfly is quite mathematically intricate and not that easy to describe. This suggests that there is some underlying structure in the chaos of Lorenz' system – something that

---

[2]Using a bit of Calculus lingo, $(x(t), y(t), z(t))$ will describe a curve in three-dimensional space, parametrized by the value of time t. Such curves are known as *trajectories* in the business. Think of a trace of vapor that a jet plane leaves behind in the sky as an apt analogy.

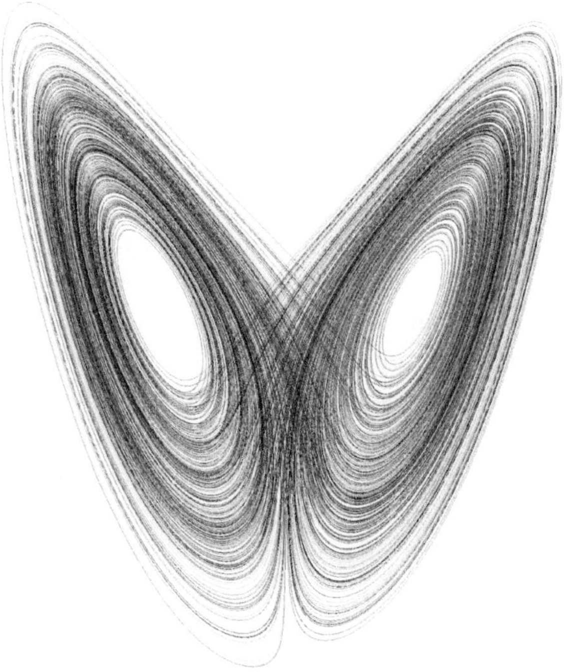

A picture of the Lorenz butterfly attractor. The actual attractor lives in three-dimensional space, if you rotated the plane of view, you would have seen that the wings of the butterfly are partly folded.

one can attempt to understand, and perhaps make predictions about, even if exact calculations are of no use.

The King of Sweden's prize

> A very small cause that escapes our notice determines a considerable effect that we cannot fail to see, and then we say that the effect is due to chance.
>
> Henri Poincaré, 1903

To me, one of the most surprising aspects of this and similar experimental discoveries of the early 1960s is that they *were* surprising. What we now know as "Chaotic Dynamics" was discovered by the French mathematical genius Henri Poincaré at the end of the 1800s. Poincaré found chaos in our very own Solar System. The actual story of this discovery is full of drama. It has been meticulously documented (I can recommend the well-researched book [2]), so I will only give you a brief summary.

Firstly, the door to finding chaos in the seemingly orderly motion of the Solar System was opened by none other than the great Isaac Newton. If you recall, Kepler's model of celestial motion was completely orderly, with all planets rigidly following elliptic orbits. In his famous oeuvre "The Principia", Newton introduced the law of gravity, which postulated that there is a force of attraction between any two physical bodies, not just celestial ones. The magnitude of this force decreases very rapidly as the bodies are moved far apart, as it is proportional to the reciprocal of (i.e. one over) the square of the distance. It is also proportional to the product of the masses of the two bodies. For two humans, the product of the masses is too small to create a significant force, so we can safely ignore Newtonian attraction as we interact with each other and focus on other types of interpersonal attraction. But replace humans with celestial bodies, and things change dramatically. Newton's introduction of the law was not without controversy, as another English scientist and a fellow member of the Royal Society, Robert Hooke, he of Hooke's Law fame, claimed that Newton "borrowed" the inverse square law of attraction from him (which is entirely possible).

How AI imagines Isaac Newton pondering gravity.

The law of gravity replaced Kepler's magic of orderly motion in elliptical orbits with a different kind of magic – a force of attraction at a distance. Newton mathematically derived Kepler's three laws of celestial motion in "The Principia" from the law of gravity using Calculus that he co-invented[3].

Consider, for instance, a pair of celestial bodies. Newton showed that if no other forces but gravity between the two of them are involved, then each will move around their common center of mass in an orbit which is shaped as a so-called *conical section*: an ellipse, a parabola, or a hyperbola[4]. If one is the Sun and the other is a planet, such as the Earth, then the center of mass of the two is pretty much the same as the center of mass of the Sun itself – the Earth is more than 300,000 times lighter than the Sun. And since the orbit of the Earth is a closed loop, it will be an ellipse in agreement with Kepler's laws. Newton thus mathematically resolved the problem of

---

[3]The authorship of the invention of Calculus caused a much more famous priority dispute between Newton and his German contemporary Gottfried Wilhelm Leibniz. It is now generally believed that the two geniuses made the discovery independently of each other.

[4]More precisely, the common center of mass will be a focal point of the conical section.

describing the motion of *two* celestial bodies in accordance with the law of gravity, sometimes known as the Two-Body Problem[5]. Parenthetically, examples of celestial bodies, comets or asteroids, which approach the Sun along parabolic or hyperbolic orbits, never to return, have since been observed.

Newton's derivation of Kepler's laws sealed the deal with the general acceptance of his own laws of motion. School children learn them to this day. But they have also opened a Pandora's box of mathematical complications. Take, for example, not two but *three* celestial bodies. For instance, consider the Sun, the Earth, and the Moon. The Moon is close enough to the Earth to be caught by its force of gravity. Its orbit is again very nearly elliptic around the common center of mass. But the gravity of the Sun affects lunar motion too – so what is the precise shape of the resulting orbit? This is not a purely academic question. It could, in principle, happen that the influence of solar gravity is disrupting the seemingly stable elliptical lunar orbit enough to cause the Moon to fly off it eventually. If one were to write a grant proposal for studying the Three-Body Problem of describing the orbits of three celestial bodies connected by Newtonian gravitation, this apocalyptic scenario could be a catchy introduction. And it is natural for mathematicians, having solved the problem for two space objects, to progress to the problem for three out of pure curiousity.

---

[5]In academia, the Two-Body Problem jokingly refers to the much harder problem of securing a position for the second member of a couple of married academic researchers (a common occurence) after the first one has obtained a tenured appointment.

My frivolous take on the Three-Body Problem as a version of the famous
"Distracted boyfriend" internet meme.

For two hundred years since Newton's solution of the Two-Body Problem, its Three-Body counterpart remained unresolved. In 1885, it became
the focus of an unusual mathematical competition. The idea came from
one of the most influential figures in European Mathematics of the time,
the Swede Gösta Mittag-Leffler. Among his many other initiatives, Mittag-
Leffler founded *Acta Matematica*, which to this day remains one of the
most prominent and prestigious mathematical journals. To promote *Acta*,
and help secure some funding for the journal, well-connected Mittag-Leffler
tried to get King Oscar II of Sweden and Norway interested in it. The King
(not a mathematician himself) became the first subscriber in 1882 and re-
ceived the very first copy of the journal.

Mittag-Leffler had another brilliant idea for getting the King involved.
He suggested to hold a mathematical competition in honor of the King's
60th birthday, which was to come in 1889. A jury, consisting of several
prominent mathematicians of the time, would propose several questions of
great scientific importance. Competitors would submit their manuscripts
to the jury anonymously by June 1, 1888. The winning work would then be

selected by the jury, and published in the "birthday" issue of *Acta Matematica*. Four questions were formulated by Mittag-Leffler's friends and mentors, Karl Weierstrass of Berlin and Charles Hermite of Paris. The most prominent of them was the Three-Body Problem[6].

Unsurprisingly, Poincaré, at that time the world's foremost young mathematical talent, became involved in the project. His entry into the competition was a manuscript with his solution of the Three-Body Problem, titled *Sur le problème des trois corps et les équations de la dynamique*. Based on an entirely new approach to studying equations of motion, it was head and shoulders above the competition. Poincaré was declared the winner of the contest, and was given a gold medal and a sum of 2,500 Swedish Kronor while his manuscript was being typeset in the *Acta*.

The prize amount was a very large sum at the time, enough to buy more than a kilogram of gold. It was, however, less than the publication cost of an issue of *Acta Matematica*, as Poincaré soon discovered. Clarifying the manuscript to one of the journal editors, the Swedish mathematician Lars Edvard Phragmén, unraveled a mistake in the reasoning. After a sleepless night, he notified Mittag-Leffler by a telegram. In a panic, the presses were stopped and the whole issue of the journal was destroyed. The printing costs which Poincaré paid out of his own pocket exceeded the prize money by 1000 Kronor.

Poincaré urgently revised his work, and a new manuscript was ready a year late, which he then expanded into a 3-volume academic monograph [15]. What he discovered while fixing the error was Chaos – the planetary motion in a Three-Body system could be chaotic. No exact description of the long-term behavior of the celestial bodies would be possible in this case. An entirely new mathematical discipline was born, known as Chaotic Dynamics. It took another 70 years for it to burst into the public consciousness after Lorenz made his numerical experiments. Which it did in a serious way. Just one example: Michael Crichton's novel *Jurassic Park* published in 1990, as well as the Hollywood movie adaptation of the same

---

[6]In fact, Weierstrass and Hermite were so optimistic that they suggested working on the general question of the motion of n celesial bodies for $n \geq 3$.

name filmed by Steven Spielberg in 1993, feature a "chaotician", Dr. Ian Malcolm.

Of lynx and hares

> A mathematical problem should be difficult in order to entice us, yet not completely inaccessible, lest it mock at our efforts.
>
> David Hilbert, the German mathematician, 1900

In fact, Poincaré discovered that in systems with only two degrees of freedom, Chaos *never* makes an appearance. The evolutionary trajectory in two dimensions can become "unbounded" – think a comet flying out of the Solar system never to be trapped by the Sun's gravitational pull again – which is usually a not very interesting situation, or else has to settle either into periodic behavior, endlessly repeating itself, or into an "equilibrium", when there is no change at all with the passage of time. This fundamental fact became known as the Poincaré-Bendixson Theorem. Ivar Bendixson was another Swedish mathematician, who made Poincaré's insight mathematically precise. Contrast this with the chaotic behavior of the trajectories of the three-dimensional Lorenz system and the intricate shape of the Lorenz butterfly attractor – none of this complexity is possible in the phenomena which can be described by the values of only two variables.

Despite the absence of Chaos, some interesting stories can be told about the evolution of two-dimensional systems. First, a real-life example of such a system with periodic behavior. It comes from ecology. In the early 20th century, scientists became increasingly interested in mathematical models of interactions between several species living in the same ecosystem. Such models are generally known as Lotka-Volterra models after Alfred J. Lotka and Vito Volterra, who independently came up with what is tellingly known as the *predator-prey model*. This model describes mathematically the interaction of two species, one of which, as you may have guessed, preys on the other[7]. These models generally predict that as time passes, the numbers of predators and prey become locked into periodic oscillations (and as one of the two populations increases, the other goes down and vice versa). This is, of course, in full agreement with the Poincaré-Bendixson Theorem: since neither of the populations can grow without a bound, periodic oscillations or being stuck at an equilibrium are the only two possible behaviors[8].

The periodic changes can take years to manifest, and estimating sizes of the populations is challenging. The model predictions might have been too hard to confirm if not for a lucky accident. Founded in 1670s, the Hudson's Bay Company held a monopoly on fur trading for several centuries in much of present-day Canada. Among the valuable furs the company bought were Canada lynx (*Lynx canadensis*) and snowshoe hare (*Lepus americanus*). Fortunately for us, and unfortunately for the snowshoe hare, Canada lynx's diet consists of basically nothing else. The lynx and the hare are a perfect example of a predator-prey ecosystem. Their numbers in the wild correlate

---

[7]Other fun examples of ecosystems with only two antagonistic species come from epidemiology. In those models, an infectious agent (a virus or a bacteria) is the predator, and humans are the prey.

[8]When one tries to model interactions of three species, the situation becomes much more complex. In particular, Chaos may appear, rendering long-term predictions useless.

tightly to the number of their furs bought by the company, and records over many decades confirm the periodic fluctuations predicted by the model.

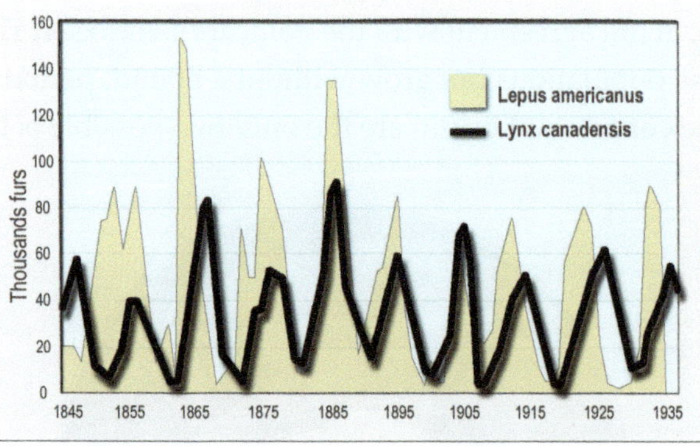

Oscillations in the number of furs bought by the Hudson's Bay Company for Canada lynx and snowshoe hare. Source: Wikipedia, by Lamiot.

One difficulty of applying the Poincaré-Bendixson Theorem to two dimensional models lies in finding all possible periodic regimes. Of course, we could just pick some initial conditions (say, the sizes of lynx and hare populations) more or less at random, run the model on the computer for some extended length of time – and then see what periodic cycle it settles into. What if the model allows for several different kinds of periodic cycles, depending on what our initial numbers were? If there are only a few of those, we could still try to identify all of them by observing the behavior of different computer simulations. But what if there are many – too many – distinct types of periodicity?

The sheer number of different types of periodic regimes predicted by the model would make it incomprehensible as a whole. We would not be able to identify all of them in practice, and would not be able to predict from the initial numbers what sort of a periodic oscillation would result. It would be best to stick to the two-dimensional systems in which the number of such cycles is limited.

In the year 1900, the illustrious German mathematician David Hilbert presented 23 mathematical problems "of the new 20th century" at the International Congress of Mathematicians. One of them – the 16th problem – was a question of finding the exact number (or at least the maximal possible number) of periodic cycles that a two-dimensional model could settle into[9], assuming that the description of the model was *algebraically simple*. The simplicity was understood as "being given by polynomial equations", similar to the Lorenz system. That example should have already taught us the lesson that models which are easy to describe may be hard to understand. And indeed, for most of the 20th century, it was not even known whether algebraically simple two-variable models always have *finitely many* possible limiting periodic behaviours. It was only in 1991/92 that mathematicians Yu. Il'yashenko and J. Écalle, working independently, proved finiteness. No bound on the maximal number is known at the time of writing – quite a challenge for the practitioners of modeling.

Chaos is a fascinating mathematical obstacle to knowledge. It seemingly devalues the precise scientific understanding of Nature. Even if an incredibly precise weather model is supported by the best measurements and is implemented on the fastest super-computer, it would still not be useful for making a forecast just a few days into the future. And yet, somehow, weather forecasting is still a thing, and we still listen to meteorologists estimates of the likelyhood of rain or shine. So despite Chaos these estimates must be useful? How? I will try to shed some light on that in the next chapter. As it turns out, Chaos is simultaneously less and more scary than commonly thought.

---

[9]Hilbert's question only concerned the cycles which are *attracting*, that is, if starting near such a cycle, a trajectory will converge to its periodic oscillation over time. They are also known as *limit cycles*, for obvious reasons.

Chapter 6

Playing dice with the Universe

> God does not play dice with the Universe
>
> ———
>
> Albert Einstein, 1926

> God tirelessly plays dice under laws which He has Himself prescribed
>
> ———
>
> Albert Einstein, 1945

Despite the evasiveness of a weather forecast on the question of whether it will rain or not the coming weekend, it will attempt to assign a numerical value to the *probability* of precipitation. It will sound something like "Precipitation 80%" which one will intuitively understand as "it is very likely that there will be rain" since certainty is 100% and 80 is pretty close to 100.

This goes well with the idea that, in practice, we often do not need to aim for an *exact* understanding of the workings of the universe to function. A general idea of how things will *likely* turn out could be good enough. Such intuitive thinking can often be put on mathematical footing in the language of Probability Theory, which attempts to quantify the likelyhood of events and outcomes. An excellent example, and a great source of homework exercises for students of Probability, is a card game of chance such as

M. Yampolsky, *The End(s) of Knowledge*, Copernicus Books,
https://doi.org/10.1007/978-3-032-01423-8_6

poker or blackjack (or even the game of dice that God may or may not play). We cannot know the cards of the opposing players in the former, or the next card in the dealer's deck in the latter, but we can quantify our chances of winning based on the incomplete information we do possess. Good players do this without thinking based on "experience" and "intuition" – but exact numerical formulas for probabilities are available. Such a probabilistic or intuitive approach to understanding the world is seemingly different from the deterministic descriptions of previous chapters. Does it lead to different scientific insights? What are its mathematical limitations?

And where does the number 80% in a weather forecast come from? 80% of what is it?

Rolling the dice

> The record of a month's roulette
> playing at Monte Carlo can
> afford us material for discussing
> the foundations of knowledge.
>
> Karl Pearson, the founder of
> Mathematical Statistics

Exact weather predictions are unreliable – worse, useless because of Chaos. So how *are* those mysterious percentages obtained in a weather forecast? What is done goes something like this: a model of the weather phenomenon is described mathematically and implemented on a computer. Think of Lorenz's experiments from the previous chapter: tiny changes in the initial values of the parameters of the model (today's measured temperature, wind direction, clouds, anything else you can think of) will likely change the forecast drastically. If one is interested in whether it will rain or not on the weekend, one might as well be tossing a coin. And a form of coin tossing is, in fact, done by meteorologists. The initial values are selected *at random* in the right ballpark. This is done many, many, many times. For each of these random choices of initial values[1] the model gives a yes/no answer to the question of whether there will be rain on Saturday. The likelyhood of rain in the forecast is the percentage of the times the simulation spits out a "yes".

By analogy with repeatedly throwing dice in a casino, this is known as *the Monte Carlo method*. It was discovered by the Polish-American mathematician Stanisław Ulam in 1946. Ulam, like many other European scientific refugees of the era, worked on the Manhattan Project, developing and perfecting first nuclear and later thermonuclear weapons. The discovery was triggered by an extended hospital stay. Likely bored out of his mind, Ulam was passing the time playing solitaire. As a bored mathematician would, he entertained himself with trying to calculate the chances that the

---

[1] It is instructive to think of it in the following way: there is a "true" set of measurements of current weather conditions. But our instruments do not work with absolute precision, so every measurement comes with a bit of random noise. Repeating the measurement, we are sampling the noise. Randomly selecting the values near the "true" value imitates this noisy error of measurement.

solitaire could be laid out successfully if the deck is shuffled at random. This is a ridiculously difficult calculation to make. Instead of simply giving up, Ulam came up with the following idea. If he played *a lot* of games, shuffled the deck well in between, and kept track of what percentage of them were successful, then he would have a good estimate of the probability of a success without having to make any calculations.

As Ulam later recalled [4]: "I immediately thought of problems of ... questions of mathematical physics, and more generally how to change processes described by certain differential equations into an equivalent form interpretable as a succession of random operations." In other words, random sampling could replace careful computations to find out the likelyhood of an event.

Soon after, Ulam described the idea in a letter to a colleague at Los Alamos, the Hungarian-American mathematical genius John von Neumann. Von Neumann was instrumental in the development of the first ever programmable digital computer – the ENIAC. Armed with Ulam's idea, the pair, joined by other colleagues such as physicists Enrico Fermi and Nick Metropolis, developed Monte Carlo computer models for complex physical processes involved in thermonuclear reactions [13]. The name of the method was suggested by Metropolis, who apparently remembered Ulam's tales of his uncle who loved to gamble in Monte Carlo casinos.

By the time Lorenz's butterfly appeared, the method was already well known. It has been used for weather prediction ever since – and not only weather prediction. Pretty much any forecasting of a multi-parameter system, be it city traffic, or nuclear explosions, is done using the Monte Carlo method.

It remains to ask whether it actually works.

It should be pretty obvious why Ulam's original idea would work for the game of solitaire. There is a finite, although very large, number of possible solitaire card layouts. While many of them lead to a dead end, some

Stanisław Ulam holding the FERMIAC – an analogue device for performing Monte Carlo simulation of diffusion of neutron particles designed by
Enrico Fermi

proportion is successful. This proportion is equal to the probability of having a "winning" layout at random. If the game is played many times, with
a careful shuffling of the deck, then the proportion of wins will converge
to this probability. This common sense statement is known as The Law of
Large Numbers.

To see how something like that could be true for a chaotic dynamical
system, let us look at a very simple abstract example – a "toy model" as
a scientist would say. The circumference (which is a fancy way of saying
length) of a circle of radius one is equal to 2π. If we take a measuring tape
and fit it around the circle, then, after one full turn, the number 2π on the
tape will cover the number 0, the origin on the tape. In that sense, the circle
can be thought of as the interval $[0, 2\pi]$ whose ends were glued together
(by convention, the interval is wrapped around the circle in the counter-
clockwise direction – that is, going from 0 to 2π left-to-right corresponds

to making one full counterclockwise turn around the circle[2]). Another turn of the tape would end by covering the origin with the number 4π, then 6π and so on.

Consider the transformation which sends the point marked x on the tape to the point marked $2x$. So, for instance, 0.25 becomes 0.5. The point π which divides the interval $[0, 2\pi]$ in half becomes 2π, which is the same point as 0. If x is greater than π, then to see where its image ends up on the circle, we have to follow the second turn of the tape. For instance, starting with 1.5π we get 3π. To get to 3π starting from zero on the measuring tape, we do one full turn, which is 2π, and then another π, ending at the point π.

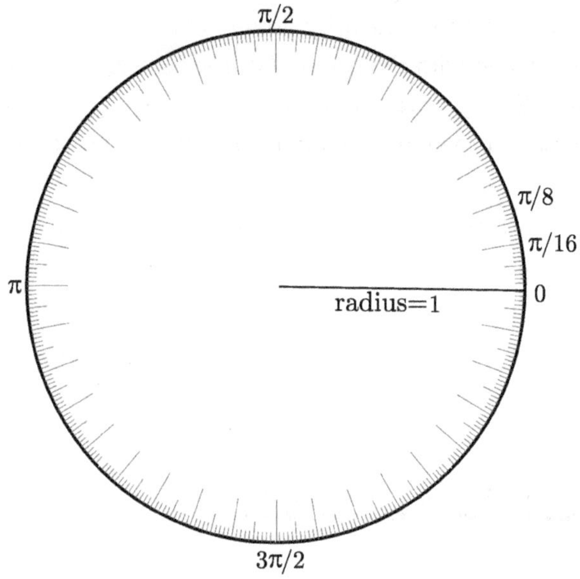

A ruler of length 2π wrapped around the circle of radius 1 in the counter-clockwise direction. Fractions of π are marked. $D(0) = 0$, $D(\pi/16) = \pi/8$, $D(\pi/2) = \pi$, $D(\pi) = 2\pi$ which brings us back to 0, $D(3\pi/2) = \pi$, and so on.

This simple rule is known as the *doubling mapping* of the circle. Let us denote the image of x by $D(x)$. If you would like a mathematical formula,

---

[2]This convention feels "wrong" as we are used to how numbers increase on the face of a clock. Its justification comes from trigonometry: if the center of the circle is placed at $(0,0)$ in the xy-plane and 0 on the tape corresponds to the point $(1,0)$ on the x-axis, then, going counterclockwise, θ on the tape covers the point $(\cos\theta, \sin\theta)$ on the circle. Oh, and clocks were designed to emulate the direction of the motion of the shadow of a sundial in the Northern hemisphere.

then $D(x) = 2x$ for $x \in [0, \pi]$, and $D(x) = 2x - 2\pi$ for $x \in [\pi, 2\pi]$. The mismatch of the two formulas at $x = \pi$ does not matter: one of them gives 2π and the other gives 0 which is the same point on the circle. The doubling mapping stretches the interval $[0, 2\pi]$ twice, doubling the distances between any two points – and then wraps it around the circle.

We will now look at $D(x)$ as the evolution rule in our toy model. Each time the clock is incremented by one unit (say a second) we will apply $D(x)$ to each point of the circle. The successive images $D(x)$, $D(D(x))$, $D(D(D(x)))$, $\cdots$ of a point x will be called its *orbit*. Surprisingly, this simple dynamics is chaotic. Suppose we change the initial point x by some tiny amount δ which can be thought of as an error of measurement. After applying the doubling mapping once, this error will double – the distance between $D(x)$ and $D(x + \delta)$ is 2δ. One more application of the doubling mapping, and the distance becomes 4δ, and so it goes, doubling at every step, until it overshoots π (then it will go down, since no two points on the circle are further apart than by π, only to increase again later on). Ten consecutive doublings would increase the error by a multiple of $2^{10} = 1024$ – rendering the discrepancy of a magnitude of one tenth of one percent at the first step into an unrecognizable end result.

Let us now designate the half-circle from 0 to π as "rain", and the half from π to 2π as "no rain" and see how well forecasting will fare in our toy model. We may get lucky for some specific initial values of x. For instance, if we start with $2\pi/7$ (rain), the doubling will first take us to $4\pi/7$ (rain), then to $8\pi/7$ (no rain), and finally to $16\pi/7$. I said "finally" since $16\pi/7$ differs from $2\pi/7$ by one full turn around the circle, so we are back to where we started. To summarize, there are only three points in the orbit of $2\pi/7$, permuted cyclically by the doubling mapping. We can make exact predictions starting from this point, as the cycle completes every three ticks of the clock. We can also make statistical predictions: starting with $2\pi/7$ we will see rain $2/3$ of the time.

But remember that we are dealing with a chaotic model. Since a truly exact initial measurement will be impossible in Nature, we should assume that there is some error – and no matter how small that error is, it will very rapidly make all predicitons useless. In the spirit of the Monte Carlo approach then, we should ask what proportion of the time an "average" orbit spends in the rain zone. Unlike the game of solitaire, there are infinitely many orbits, so averaging is not straightforward. However, with some higher level mathematics, it is possible to show that if we were to select a point on the circle at random – say closing our eyes, and pointing blindly – then with probability 1 it will produce an orbit which fills the circle uniformly, and spends a proportion of time L in each arc of length L. If it is true for "most" orbits, it is also true on average, so a Monte Carlo simulation would give us a 50% chance of rain since our rain zone covers a half-circle.

Losing at Monte Carlo

> Gambling: A tax on people who are bad at math.
>
> A proverb

Toy models are extremely useful in developing our intuition on the behavior of complex dynamical systems we may encounter in Nature. The Wright brothers would hardly know how to construct a flying machine if not for the centuries long history of toy airplanes which helped develop both the science and the intuition of flying. Unbelievably, the oldest known toy airplane dates to 200BC! It was unearthed in 1898, in an Egyptian archeological excavation. A very modern-looking propeller-driven model airplane named the *Planophore* was built by the French inventor Alphonse Pénaud in 1871. As kids, the Wright brothers were given a toy helicopter – another one of Pénaud's designs – by their father. They credited it for their early interest in flight.

Since Lorenz's experiments, mathematicians have put enormous efforts into the study of toy models of Chaos. In fact, most of these efforts since Lorenz's time went into understanding one particular *family* of toy models, which combines simplicity of definition with astonishing complexity of behavior. It is again a discrete-time model with one degree of freedom – the position of a point in the interval $[-1, 1]$. One tick of the clock moves this point to its image by a quadratic function. If we were to plot this function, we would see a parabola whose tip is over (or under) $x = 0$, at the middle of the interval. The endpoints -1 and 1 are both sent to 1, and the image of 0 is some point a between -1 and 1 – this guarantees that the whole image fits inside the original interval $[-1, 1]$.

Visually, the interval is folded twice, and laid inside itself starting from the right end point. Algebraically, the rule can be expressed as[3]

$$x \longrightarrow f_a(x) = (1 - a)x^2 + a.$$

For illustration, I have drawn the graphs of functions f_a for several different a's in the figure below – but there is really not very much to see there. The

---

[3]Amusingly, the quadratic polynomial $f_{-1} = 2x^2 - 1$ is the doubling mapping $D(\theta)$ in disguise. If we plug $\cos\theta$ into f_{-1}, then $f_{-1}(\cos\theta) = \cos 2\theta$ according to a standard trigonometric identity – so the action of f_{-1} on values of $x \in [-1, 1]$ transforms into $D(\theta)$ for $x = \cos\theta$. Polynomials which transform into multiplication by an integer $m > 1$ in this fashion were studied in the 1800s by the Russian mathematician Pafnuty Chebyshev, and are named after him. Another fun fact is that Ulam and von Neumann themselves used f_{-1} as a test for Monte Carlo simulations.

graphs are all parabolas, the only difference between them is the vertical position of their tips (whose heights are the corresponding values of a).

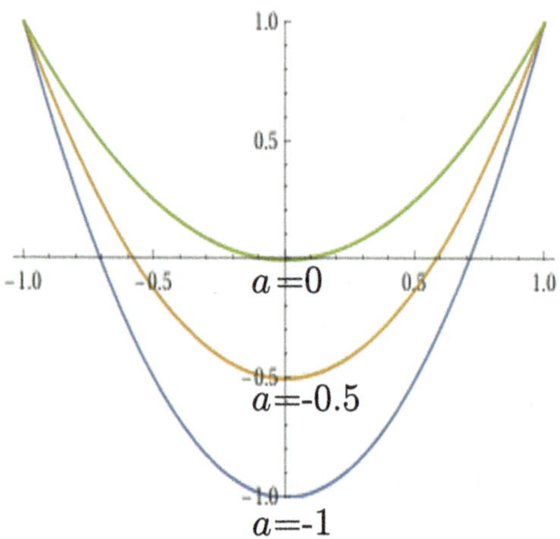

The graphs of the parabolas f_a for a few different values of a.

And yet, varying the value of a, one can obtain wildly different statistics for the orbits of f_a. You can take my word for it and skip the next little section, to just past the next figure – it is quite mathy – or dive into it if you are mathematically inclined.

To get some idea of just how much the statistics of the orbits may vary depending on a, look at the right-hand side of the figure below. It is an attempt to visualize the limiting behavior of typical orbits of f_a. The value of a is plotted on the horizontal axis, from -1 to 1. On the vertical interval corresponding to each value of a, is a plot (or at least an attempt at one) of the possible limits of the orbit of a randomly chosen point x under successive iteration by that particular f_a.

For instance, when the values of a are close to 1, the orbits of all points converge to 1 which is left stationary by f_a. This is indicated by plotting a single point at height 1 over each such value of a. Of course, the statistical analysis of orbits is laughably easy for maps like this: no matter what initial

x is chosen, after a while the orbit will settle near 1 with certainty. The probability of any other outcome is zero.

If we move left past the value $a = 0.5$, then there is still only one possible limit for a typical orbit, but it is no longer $x = 1$. We can see that the previously horizontal curve bends down as we move left. As we reach $a = -0.5$, a major change happens: past this point, there is not one but two limit points for a typical orbit. These two points are interchanged by f_a so the limiting behavior of an orbit is not stationary but periodic, with period 2. The curve "bifurcates" at $a = -0.5$: it splits into two curves and going further left there are two values corresponding to each a. The statistical analysis is still easy: an orbits' time will be equally split between the two points of the periodic cycle.

Going further left, we see another bifurcation: a cycle of period 2 is replaced by a cylce of period 4. In fact, there are infinitely many such bifurcations, but they are hard to discern since soon after the picture becomes a seemingly intractable mess, with the limiting behavior of orbits too difficult to plot. The dynamics actually becomes chaotic.

As a side remark, the picture itself, known as a "bifurcation diagram" has a fascinating fractal structure. There are some self-similarities in it, for instance, in the way that the "period-doubling" bifurcations appear. You may be amused to learn that the "renormalization revolution" in mathematics originated from the study of this bifurcation diagram. In the late 1970s mathematical physicists Mitchell Feigenbaum and, independently, Pierre Coullet and Charles Tresser, gave a visionary explanation of this self-similarity. By analogy with renormalization ideas which by then conquered physics, they conjectured that the dynamics of f_a can be renormalized. Successive renormalizations of f_a would then be expected to converge to a self-similar evolution rule. This idea took decades to be fully verified, and my own introduction to the subject of renormalization began with its study.

But I have digressed. I only wanted to convince you that the statistical behavior of orbits of the very simple-looking evolution rule f_a can be quite complicated as well as interesting to study, with enough variety to make the family f_a a perfect test bed for the Monte Carlo method.

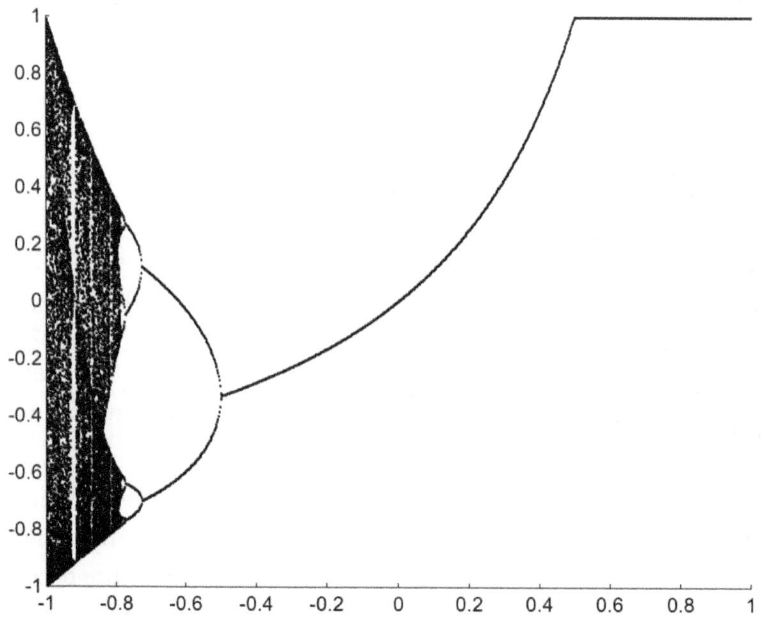

The bifurcation diagram of the family f_a. The values of a are marked horizontally; on the vertical segment over each a the possible limits of a typical orbit are indicated.

A couple of years ago, together with a Chilean colleague Cristobal Rojas, we decided to study the question whether the Monte Carlo method would always "work" for toy models in the family f_a regardless of the value of the parameter a. What we found, and published in [16], was both unexpected and discouraging. As we know, for the Monte Carlo approach to make sense, there should be a nice statistical distribution for the orbits of the dynamical system – the statistical distribution which we attempt to sample. This was thought to be the main danger for the technique. Luckily, it is not a problem for these particular toy models – for a typical value[4] of a, for any randomly chosen[5] starting point x, the long-term statistical behavior of the orbit of x will be the same. The proof of this was given by my PhD advisor, Mikhail Lyubich [11]. That would be ideal for Monte Carlo modeling. But shockingly, as we discovered, for many such maps, the statistical distribution itself cannot be computed.

---

[4]For instance, if a is chosen at random, this will be true with probability 1.

[5]Again, with probability 1.

No algorithm which would correctly tell the likelihood of an orbit lingering in a "rainy" zone could possibly exist for such models. Remarkably, Turing's Halting Problem is again the culprit. Since no numerical technique would work for such maps, the Monte Carlo approach would fail[6].

The jury is out whether this phenomenon replicates itself in multi-dimensional real-world models, but our proofs make it more than likely. Worse than that, it is likely that with non-zero probability, one will encounter dynamical systems for which the long-term statistics of the orbits is not computable in practice – either no algorithm exists at all, or any algorithm that may exist would require an unrealistic amount of running time.

Let us take a step back. The mere fact that a nice statistical distribution for the orbits of a dynamical system exists should not be taken for granted. It took several decades to determine whether this is true for the toy model family f_a.

It is not a given that all (well, all *typical*) orbits of a dynamical system will follow the same statistics. The fun picture below illustrates the paper [3] (used with permission). The model it describes is two-dimensional: its orbits live in a cylinder. There are two possible limiting statistical distributions. One of them is on the upper circular boundary of the cylinder, and the other one is on the lower one. These are two different versions of the future that the model describes, and they are quite distinct. Which one we get depends on the starting point of the orbit. The black and grey colors in the picture of the cylinder are an attempt to distinguish the orbits corresponding to the lower circle from the orbits corresponding to the upper one. But any such attempt at coloring is ultimately futile – no matter how much we zoom into the drawing, we will still see both colors present. The two statistical behaviors are perfectly intertwined.

---

⁶It would fail sneakily, not letting us suspect that we are getting wrong predictions. This is how non-computability must work in practice, so we could not use a brute force search for the correct answer by discarding the wrong ones.

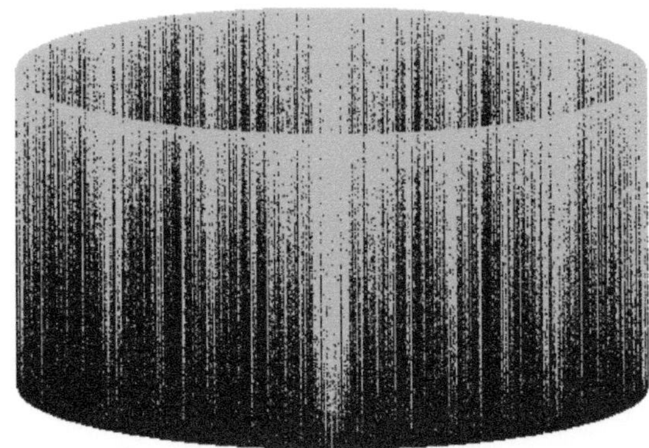

In itself, such a phenomenon would not be an obstacle for making predictions. Using the weather analogy, a model with two statistical outcomes would describe two different *scenarious* by which a weather sysytem would develop. By testing some initial values, one could hope to figure out the probability of rain in each of these scenarios as well as the likelyhood of either scenario unfolding.

It has long been expected in mathematicians circles that, for a typical dynamical system, only finitely many different statistical outcomes exist for the distribution of orbits. Named the Palis Program, after the illustrious Brazilian mathematician Jacob Palis [**14**], this has been the guiding light for the research activities of the last several decades. However, the model one-dimensional family f_a remains the only *proven* instance of the conjecture. And as we learn more about the intricacies of dynamical systems with more than one degree of freedom, doubt has started to creep in.

In the vision suggested by the Palis Program, given a physical system to study, we can randomly pick a point, start following the orbit, and keep track of the statistical distribution of its position. The longer we will iterate, the more the statistical picture will clarify itself, converging to one of finitely many possible scenarios. But what if, for a typical orbit, the picture does not become clearer as we keep iterating? What if it gets murkier instead? A French colleague, Pierre Berger, has studied an extreme situation, when, as we follow the orbits, with each tick of the clock the number of possible statistical scenarios *grows*. In 1941, the famous Argentinian writer Jorge Luis Borges wrote a short story "The Garden of Forking Paths". Deeply philosophical, as much of his writing, it conveys that the future is not a straight line. Rather, every possible outcome for each event creates several different versions of the future, which all coexist, creating a tree of bifurcating paths. The systems that Berger has described behave in this fashion – following the orbits leads to a sequence of bifurcating statistical predictions of the future, all of which coexist at the same time. The longer we observe the dynamics, the more possibilities emerge, with the predictions getting more and more complex, rather than settling down to a finite collections of scenarios.

As with our non-computability result with Rojas, the jury is still out whether the "forking paths" scenario described by Berger can be encountered in physical applications with a positive probability. But I would not want to bet against it. Chaotic dynamics may turn out to be truly unpredictable, after all.

TL;DR

Endings to be useful must be
inconclusive

Samuel R. Delaney, American
Sci-Fi writer

It is a dangerous thing to summarize the content of a book in its conclusion – the reader may be tempted to skip reading the stuff in the middle. Nonetheless, here goes.

We often imagine knowledge as an ever-expanding frontier, a limitless terrain awaiting exploration. This is not the case – it is fenced in by mathematical boundaries which sharply limit the scientific quest to explain and understand nature.

A basic limitation of knowledge that raises its ugly head again and again in the book is what is informally known in mathematical sciences as "the curse of dimensionality". Phenomena which require many variables to describe are truly unknowable. And by "many" I do not actually mean that many. Depending on the context, the number is usually somewhere around four or five. This limitation is based in mathematics and is therefore immovable and universal. It applies to biological and artificial brains alike. If something does not have a simple description, it will never be understood.

© The Author(s), under exclusive license to Springer Nature Switzerland AG 2025
M. Yampolsky, *The End(s) of Knowledge*, Copernicus Books,
https://doi.org/10.1007/978-3-032-01423-8_7

Mathematical models built on many variables are worse than useless. They are fundamentally anti-scientific as they can be used to explain away anything.

The existence of a simple description is no guarantee that it will be found. It is possible to describe mathematically the intutive process of refining the empirical data to discard irrelevant parameters and focus only on what is important. It underlies much of modern physics, under the name of "renormalization". It is also a fundamental principle of cognition. Yet, there is no gurantee of success. Worse, there is a mathematical guarantee that a truly general approach for distilling the simplest possible explanations of empirical data will never be found.

Finding a simple, low-dimensional, law of nature is a scientific miracle. But it may not result in useful predictions. The "noise", omnipresent in the real-world data, may be amplified by a mathematical model to an extreme extent, making the outcome no better than a random guess. Known as Chaos, this mathematical fact seems to negate the very purpose of searching for scientific knowledge.

As a remedy to Chaos, scientists have developed a statistical approach to modeling, aiming to describe the likelyhood of all possible future outcomes, rather than predict the exact future. Here, the curse of dimensionality strikes again, as the description of all possible statistical outcomes may itself require too many parameters – many more than the mathematical model itself has, and too many to be understood.

So even concise and simple-looking mathematical laws of nature can be hiding overwhelming complexity which defies knowledge.

Last, but not least, since (and even prior to) the advent of computers, scientists have been looking into the mathematical limitations of using them for the modeling of nature. These turn out to be quite severe, complicating the quest for knowledge even further. Even a well-behaved low-dimensional law of nature may lead to an impossible computational problem.

In light of the dire picture painted above, one is left to wonder why science works at all. The same question can be asked about cognition, which "unconsciously" solves the same basic problem of comprehending the workings of the world, and faces the same mathematical obstacles as science. One cannot help feeling that the universe "wants" to be understood.

Many great scientific minds of the past have seen this amenability of the universe to human understanding as a proof of its divine – or, in a modern jargon, intelligent – design. The implied meanings of "divine" and "intelligent" are subtly different in this context, but the sense that the world was custom built (for us? for what comes after us?) they convey is roughly the same. The half-joking half-serious references to us "living in a simulation", or in an "aquarium" have a similar flavor. Even without a specific design, the principles on which the universe was built may incorporate cognitive success. This thought is also not new and has been somewhat explored in the literature, however, such ideas cut too close to issues of personal religious belief to ever be discussed dispassionately – which makes it difficult to discuss them scientifically.

As a rebuttal, a form of survivor's bias is often asserted: science claims success where it can, while unknowable phenomena are simply ignored as "noise". Similarly, human cognition recognizes patterns and "laws" where they can be found. Thus it is not that those are universally successful, they are only universally successful in the (possibly, narrow) range where they can succeed.

One can retort, of course, that the range of cognitive success is vast enough to allow for the evolution of intelligent life which has managed to develop some very advanced science. This can be countered by saying that ours is just one of an infinite number of possible universes, and that in this particular one the stars have aligned to allow enough understanding to emerge for the development of civilization. This is also a survivor's bias argument, but for universes.

My purpose in writing this book was not to take sides in the above debate, but to highlight both the profound mathematical obstacles in search for knowledge, and my sense of awe at the success of science in spite of them. As I was writing, I could not help but wonder why so much of my own research and that of my colleagues is ultimately about the very limits of research. The answer, it seems, is that this is where the action is – the liveliest scientific debates and the most profound revelations emerge when certainty falters. Sometimes, they lead to a new scientific synthesis, sometimes to an end, a true boundary of understanding. Mapping out those boundaries is valuable in its own right as it shapes the direction of our efforts. We cannot always know what will remain unknowable, but we can have fun trying – I do.

Bibliography

[1] M.F. Barnsley. *Fractals Everywhere*. Academic Press, San Diego, 1988.

[2] J. Barrow-Green. *Poincaré and the Three Body Problem*. History of Mathematics. American Mathematical Society, 1996.

[3] A. Bonifant and J. Milnor. Schwarzian derivatives and cylinder maps. In *Holomorphic dynamics and renormalization*, volume 53 of *Fields Inst. Commun.*, pages 1–21. Amer. Math. Soc., Providence, RI, 2008.

[4] R. Eckhardt. Stan Ulam, John von Neumann, and the Monte Carlo method. *Los Alamos Science. Special issue*, 1987.

[5] G. Grah, R. Wehner, and B. Ronacher. Desert ants do not acquire and use a three-dimensional global vector. *Front Zool.*, 4(12), 2007.

[6] B.J. He. Scale-free brain activity: past, present, and future. *Trends Cogn Sci*, 18:480–487, 2014.

[7] J. Ioannidis. Why most published research findings are false. *PLoS Med.*, 2, 2005.

[8] Arthur Koestler. *The sleepwalkers: a history of man's changing vision of the universe*. Macmillan, New York, NY, 1959.

[9] E. N. Lorenz. Deterministic nonperiodic flow. *J. Atmos. Sci.*, 20:130–141, 1963.

[10] E.N. Lorenz. *The Essence of Chaos*. U. Washington Press, Seattle, 1993.

[11] M. Lyubich. Almost every real quadratic map is either regular or stochastic. *Ann. of Math. (2)*, 156(1):1–78, 2002.

[12] P. Mehta and D. Schwab. An exact mapping between the variational Renormalization Group and Deep Learning. 2014.

[13] N. Metropolis. The beginning of the Monte Carlo method. *Los Alamos Science. Special issue*, 1987.

[14] J. Palis. A global view of dynamics and a conjecture on the denseness of finitude of attractors. *Astérisque*, 261:339–351, 2000.

[15] H. Poincaré. *Les méthodes nouvelles de la mécanique céleste. Tomes 1-3*. Gauthier-Villars, Paris, 1892-1899.

M. Yampolsky, *The End(s) of Knowledge*, Copernicus Books,
https://doi.org/10.1007/978-3-032-01423-8

[16] C. Rojas and M. Yampolsky. How to lose at Monte Carlo: a simple dynamical system whose typical statistical behavior is non computable. In *STOC 2020: Proceedings of the 52nd Annual ACM SIGACT Symposium on Theory of Computing*, pages 1066–1072, 2021.

[17] Hannes Schulz and Sven Behnke. Deep learning – layer-wise learning of feature hierarchies. *Künstliche Intell.*, 26(4):357–363, 2012.

[18] A. Turing. Solvable and unsolvable problems. *Science News*, 31:7–23, 1954.

[19] A. M. Turing. On computable numbers, with an application to the Entscheidungsproblem. *Proceedings, London Mathematical Society*, pages 230–265, 1936.

[20] N. Vereshchagin and P. Vitányi. Kolmogorov's structure functions and model selection. *Information Theory, IEEE Transactions on*, 50:3265–3290, 01 2005.

[21] S. Weinberger. *Computers, Rigidity, and Moduli: The Large-Scale Fractal Geometry of Riemannian Moduli Space*. Princeton University Press, 2005.

[22] E. P. Wigner. The unreasonable effectiveness of mathematics in the natural sciences. *Communications on Pure and Applied Mathematics*, 13:1–14, 1960.

[23] Peter Woit. *Not Even Wrong: The Failure of String Theory and the Continuing Challenge to Unify the Laws of Physics*. Random House, 2011.